THE CELLULAR SLIME MOLDS

The Cellular Slime Molds

SECOND EDITION

REVISED AND AUGMENTED

JOHN TYLER BONNER

PRINCETON, NEW JERSEY

PRINCETON UNIVERSITY PRESS 1967

Preface to the First Edition

THE purpose of this monograph is twofold. First, an attempt has been made to give a comprehensive survey of all the different known aspects of the biology of the cellular slime molds. There has been a growing interest among experimental biologists working in widely different areas on these organisms, and therefore not only is the present status of the experimental knowledge stressed, but also life histories and relationships with other organisms, so that as complete a perspective as possible may be obtained. The field is yet young and the total amount of work done thus far is small, as the complete cellular slime mold bibliography in the back of the book will show. For this reason a thin volume is still possible, and it is hoped that in the years to come it will continue to serve as a useful summary of all the work done before 1959. But it must be admitted that this is not a mere summary of the field, for even though an effort has been made to include all the important facts, this is primarily an interpretation of these facts.

The cause of this may be found partly in human nature and partly in the second purpose of the book. The main reason that various workers have devoted so much of their energies to experimental studies of the cellular slime molds is that they are considered to be particularly useful organisms in the study of development. Therefore, inevitably, if one surveys and analyzes the literature, as has been done here, one hopes that by looking at all the known facts together, some new insights into the mechanism of development will appear. Writing this book has been of considerable value to me for just this reason, and I hope that it will serve as a useful stepping stone for the research of others in the future.

I should like to take this opportunity to thank Professor K. B. Raper, Dr. B. M. Shaffer and Professor C. H. Wad-

PREFACE

dington for their kindness in reading the manuscript and
providing many helpful suggestions and criticisms. Also I
should like to acknowledge my indebtedness to the National
Science Foundation and the Eugene Higgins Trust of Princeton University for financial assistance provided during the
preparation of this book and during the current experimental
work from our laboratory presented in this volume.

Finally, I am grateful to the following individuals for
permission to use their illustrations: Drs. E. H. Mercer,
K. B. Raper, B. M. Shaffer, and M. Sussman. I also wish to
thank Miss Kathleen Dodge for her original drawings.

Preface to the Second Edition

IN THE six years since the first edition was published the number of references on the cellular slime molds has jumped from 124 to 341. It has meant that in preparing this edition more than half of the book has been totally rewritten. Furthermore it is no longer possible to make a brief summary of the entire field and have it include all the facts. More than before this is a severe editing of the facts. It is primarily a discussion of the development of the cellular slime molds. The only aspect that is complete or encyclopedic in any sense is the bibliography.

From comments on the first edition it is clear that one of its uses was to introduce students and research workers to these organisms. It was a convenient survey of experimental problems in the cellular slime molds. If anything this aspect has been emphasized in this edition and, for instance, in the chapter on growth, practical information on laboratory methods are included. It is hoped that this book will serve as a useful stepping stone for any individual who wishes to engage in research on the slime molds.

As in the first edition I should again like to acknowledge my indebtedness to the National Science Foundation and the Eugene Higgins Trust of Princeton University for financial assistance provided during the preparation of this book and during the current experimental work from our laboratory presented in this volume.

I would also like to thank the following individuals for their most helpful critical reading of the manuscript: Mr. C. Ceccarini, Mr. E. G. Horn, Dr. L. S. Olive, and Dr. B. M. Shaffer. I am grateful to the following individuals for permission to use their illustrations: Dr. D. W. Francis, Dr. G. Gerisch, Dr. H. R. Hohl, Dr. T. M. Konijn, Dr. K. B. Raper, Dr. E. W. Samuel, and Dr. B. M. Shaffer.

PREFACE

Finally, I would like to express my deep appreciation to Mrs.
Anne Tittler and Mrs. Rogene Gillmor, who combined run-
ning our full-time experimental program with helping on the
bibliography of this book. I also thank Mrs. Gillmor for the
new, original line drawings.

August 1966   J.T.B.
Margaree Harbour
Cape Breton
Nova Scotia

· viii ·

Contents

CONTENTS

THE CELLULAR SLIME MOLDS

I. Aggregation Organisms

1. Slime molds

WHEN Brefeld made the first positive identification of a member of the Acrasiales in 1869, he mistakenly thought the cells fused to form a true plasmodium. This error was not corrected until 1880 when van Tieghem demonstrated that the cells remained uninucleate at all times; to underscore the fact that there was no plasmodial syncytium in the Acrasiales, the aggregated cellular mass was called a "pseudoplasmodium." As the term implies, there was originally a strong belief in the close affinity of Myxomycetes and Acrasiales, but it has become increasingly clear, as our knowledge has advanced in recent years, that there is probably no phylogenetic relation between the two at all. They differ in so many particulars that to associate them, except by remote analogy, seems at the moment out of the question. The only feature that they have in common, and this is a source of infinite confusion, is the fact that they both are called "slime molds." There are still a number of textbooks of elementary biology which, because of this general term, completely confuse the two, and it is not uncommon to find an illustration of a member of the Acrasiales with a corresponding text description of a Myxomycete. To the discerning eye of a student of these lower groups this is an unpardonable error, equal to confusing a nematode and an annelid because they both are "worms."

To separate the two unrelated slime molds one may simply use the terms Myxomycetes or Myxogastrales for one and Acrasiales for the other. If common names are desired, the former have been called "plasmodial" or "true slime molds," and the Acrasiales have been called "simple slime molds" by K. B. Raper and "amoeboid slime molds" by myself. However, I am inclined to think that the most useful term is "cellular slime molds," the name suggested by B. M. Shaffer.

To add to the confusion there are three other groups of organisms that are often associated with the slime molds already discussed, and all five groups may be conveniently classified in the following fashion:

> Mycetozoa
> > Labyrinthulales
> > Plasmodiophorales
> > Myxomycetales (or Myxomycetes or
> > > Myxogastrales)
> > Acrasiales (or Acrasina)[1]
> > Protosteliales (or Protostelida)

The logic of this grouping is that all five groups contain primitive colonial organisms that have both fungal and animal-like characters and all are to some extent slimy. Aside from these doubtful binding qualities, their relation to one another is uncertain. They are placed together because they represent five anomalous groups of unknown origin.

2. Labyrinthulales

This group is composed of one genus, *Labyrinthula*, and a mere handful of species. They also are parasites and are of considerable ecological importance, for as Renn[2] showed, they were responsible for the wasting disease of eel grass which did so much to change the whole aspect of our coasts some thirty years ago.

These marine parasites infect not only eel grass but many algae as well. The vegetative cell is spindle-shaped and produces a projection of material from each end, much as a developing nerve cell of an animal produces its axone. These

[1] Botanists and zoologists differ on the ending of names of orders, but since these organisms are neither plants nor animals the distinction is unimportant and the names are interchangeable.

[2] *Nature 135* (1935) : 544.

projections of *Labyrinthula* form a network, and the uninucleate cells proceed to glide up and down it, travelling only on the tracks they have previously laid down. This pattern of growth has been given the unfortunate name of "net plasmodium." It does indeed form a net, but since the cells are at all times uninucleate, it is in no way a plasmodium. The spindle cells occasionally aggregate into a dense mass, and each cell becomes encapsulated into a spore, but again without any syncytium. Therefore, these now represent true aggregation organisms, in a sense, although the aggregation is of such a crude and primitive nature that it fails to provide the opportunities for experimental analysis found in the Acrasiales.

Until recently there had been no indication of any flagellated cells, nor any evidence of sexuality. Independently, Hollande and Enjumet[3] and Watson[4] discovered the presence of biflagellate swarmers, but there still has been no demonstration of sexuality in the life cycle of these organisms (Fig. 1).

The phylogeny of *Labyrinthula* represents a special problem, and G. M. Smith[5] excludes it from his general slime mold category, Myxomycophyta, on the basis that it shows similarities to the golden alga *Chlorarachnion*. The possibility that Smith is correct in the conjecture that *Labyrinthula* is an alga which has lost its photosynthetic pigments upon becoming parasitic, is certainly reasonable and not disputed. I would, however, not exclude it from the group of slime molds on this basis, for we are assuming that the heterogeneous Mycetozoa are very likely polyphyletic.

3. Plasmodiophorales

The Plasmodiophorales are a group of organisms which

[3] *Ann. Sci. Nat. Zool.* 11ᵉ Series, *17* (1955) : 357-368.
[4] Ph.D. Thesis, University of Wisconsin (1957).
[5] *Cryptogamic Botany*, Vol. 1, 2nd edn. McGraw-Hill, New York, 1955.

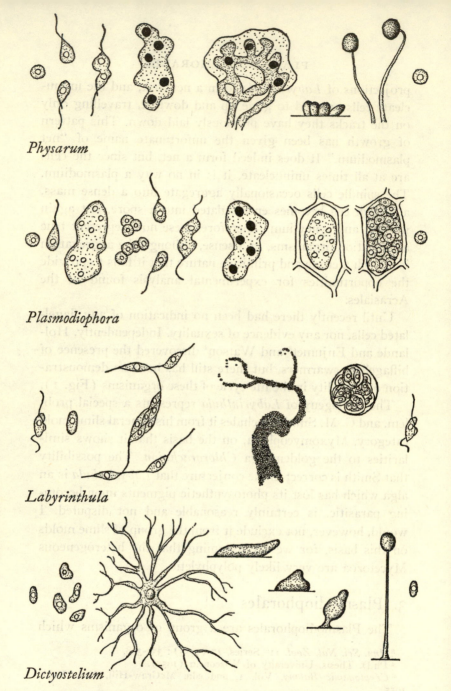

Physarum

Plasmodiophora

Labyrinthula

Dictyostelium

Fig. 1. The life cycles of representatives of the four groups of Mycetozoa. From top to bottom: Myxomycetales, Plasmodiophorales, Labyrinthulales, and Acrasiales. In the first two, where sexuality is well established, diploid nuclei are indicated by solid black dots while haploid nuclei are white. Also the first two are the only ones showing a plasmodial stage. In each the cycle begins (left) and ends (right) with a unicellular spore.

have been of interest principally to the plant pathologist, for all the known species, of which there are some thirty, are plant parasites. They are most noticeably destructive among cabbages, and the clubroot disease of cabbage was a great scourge some years back; now that effective methods of control are known, it is no longer of importance.

As will be seen from an examination of Karling's[6] volume on Plasmodiophorales, there is some variation in the cycle of different members of the group and even some variation in the interpretation of these cycles by different workers. Here I shall confine the discussion to Cook and Schwartz's[7] description of *Plasmodiophora brassicae,* which is the specific agent of the clubroot disease of cabbage (Fig. 1).

The spore of the mold germinates to produce a biflagellate swarmer, with one flagellum shorter than the other. As a matter of fact, the paired nature of the flagella was first discovered in the Plasmodiophorales in 1934 by Ledingham,[8] and only subsequently was the small second flagellum observed in the Myxomycetes. This swarmer penetrates the root of a cabbage seedling and becomes, inside a cell of the plant, a myxamoeba by reabsorbing its flagella. The swarmer (and also the myxamoeba) is haploid; therefore the plasmodium which results from mitotic divisions of the original nucleus of the myxamoeba is haploid and has no counterpart in the life cycle of Myxomycetes. The haploid plasmodium is minute, not exceeding thirty nuclei, and when it stops growing it cleaves, cutting off a number of haploid, uninucleate cells. Each of these is in fact a gametangium; by division each produces four or eight flagellated gametes. The gametes probably fuse to produce a diploid myxamoeba, and this myxamoeba, again by mitotic divisions, will produce a plasmodium which in this case will be diploid, although there is

[6] *The Plasmodiophorales.* New York, 1942.
[7] *Phil. Trans. Roy. Soc. London, Ser. B., 281* (1930): 283-314.
[8] *Nature, 133* (1934): 534.

still some doubt concerning this diploidy. Either at the uninucleate stage or after a few nuclear divisions the parasite becomes highly invasive and bores its way from one cell to another, in this way entering the cambium of the cabbage, passing to various parts of the root, and finally wandering out into the cortex. The plasmodium now becomes large and produces a corresponding effect in the cabbage root, which thickens into the club-like swellings. It is suspected that the plasmodium eventually undergoes meiosis and progressive cleavage to form numerous haploid spores.

Concerning the relationship of the Myxomycetes and the Plasmodiophorales, it should be noted that they are both of flagellate origin, but this is also true of the majority of colonial animals and plants. It was the old traditional view that the two groups were closely associated, one being parasitic and the other, free-living. However, with increased knowledge it became obvious that they differ in so many ways, such as the cytological details (e.g. the "cruciform" division charactertistic of the Plasmodiophorales mitosis) and the haploid plasmodium of the Plasmodiophorales, that the modern trend has been to consider the Myxomycetes a discrete group, with the Plasmodiophorales related to the lower filamentous fungi or Phycomycetes. This is based largely on similarities between Plasmodiophorales and Chytrids, the details of which would be superfluous in this discussion. There is a good possibility that each of these three groups, the Myxomycetes, the Plasmodiophorales, and the Chytrids, arose independently from flagellate ancestry. Perhaps all we can assert with confidence is that the origin and relationship of these groups is doubtful.

4. Myxomycetales

The Myxomycetes include some 400 species. Their fruiting bodies are macroscopic and easily recognizable on decaying wood or leaves. Most of the species bear their spores inside a sheath or peridium (Endosporeae), although individuals of the interesting genus *Ceratiomyxa* bear the spores singly on small papillae rising from the body of the mold (Exosporeae).

The life cycles of Myxomycetes were not completely understood until the work of Wilson and Cadman,[9] who were the first to clarify unequivocally the sexual nature of plasmodium formation. From their work and the work of others it is now possible to present a generalized life cycle that probably applies to a number of the members of the group (Fig. 1. See Alexopoulos[10] for a recent review).

The delicately sculptured spore germinates to liberate a haploid swarmer. In some cases the spore liberates four such swarmers, and in others a single swarmer may divide into four daughter swarmers. Either immediately after germination or after cell divisions, if they occur, the cells sprout one or two flagella. These flagella may subsequently be reabsorbed with the return of the amoeboid condition. The important point is that these cells, at either the flagellated or amoeboid stage, may serve as gametes and fuse in pairs. The resulting diploid cell is the fertilized "egg" that gives rise to the plasmodium, and now there follows a series of nuclear divisions without corresponding cell cleavages, and the protoplasmic mass begins a great period of expansion.

During this rapid growth the young plasmodium acts like a large amoeba, engulfing bacteria and other organic particles. Under the proper environment of nutriment, temperature, and moisture, the size of the plasmodium may increase to a few inches or more in diameter, and it is a common observation to see on a decayed stump in a wet forest a glistening viscous mass of slime, frequently a brilliant yellow. If adverse conditions should suddenly arise, the plasmodium can con-

[9] *Bull. Trans. Roy. Soc. Edinburgh 55* (1928) : 555-608.
[10] *Botan. Rev. 29* (1963) : 1-78.

tract into a thickened, hardened mass. This so-called sclerotium may be stored in a dry condition for some time, but it will soften and release a viable plasmodium after being replaced in a favorable environment.

Sporangium formation is a more organized affair. The protoplasm tends to concentrate in a limited region and there becomes cut up into a series of compartments; each of these rises into a small bleb which will develop into a sporangium. In stalked forms the blebs rise into the air. The protoplasm flows upward, depositing centrally a stalk of non-living material. After the bulk of the protoplasm has reached its apical position, an outside wall or peridium is secreted. A series of furrows forms, so that ultimately each nucleus is isolated in a block of protoplasm. Either before or during this process the nuclei undergo meiosis, so that the final isolated nuclei are haploid; each of these cells then secretes a hard wall to become a resistant spore. In some species a material is secreted in the larger cracks resulting from progressive cleavage and hardens to form a thread-like capillitium. Depending on the species, this capillitium may be delicately sculptured as well as hygroscopic, thereby helping the process of spore dispersal.

The variety of shapes of the fruiting bodies is great, and it is used as the basis of species classification. In some species the fruiting bodies have no stalk but are merely rounded or flattened masses projecting from the substratum. The majority are stalked, but the stalk may be single or branched. The great variation comes in the details of the structure of the stalk and the capillitium, the shape of the peridium, and the sculpturing of the spores.

It has been shown by numerous workers (see Alexopoulos[11]) that separate zygotes or plasmodia may fuse or that a whole cycle may be completed from a single zygote. In those

[11] *Ibid.*

cases where there is fusion of separate zygotes, one has a rudimentary kind of aggregation.

5. Acrasiales

My intention in this section is to give only a brief picture of the Acrasiales for purposes of comparison with the other slime molds. At the moment about two dozen species of the Acrasiales are known, all of which are free-living in the soil or are found on dung. The germinating spores liberate a single uninucleate amoeba; there are no flagellated cells. The amoebae feed on bacteria by phagocytosis and repeatedly divide by mitosis, each daughter cell remaining uninucleate and free and independent of the other cells. When a large number of such separate amoebae have accumulated, they will stream together to central collection points to form a cell mass or pseudoplasmodium. This concentration of cells involves no fusion of protoplasts, but the uninucleate character of the cells is essentially maintained following aggregation and during all the other phases of the life cycle (Fig. 1). The qualification necessary for this statement is the interesting observation of Huffman, Kahn, and Olive (1962) and Huffman and Olive (1964) that cells will form temporary anastamoses.

One important point here is that there is a natural separation between the feeding stages and the morphogenetic stages. Feeding will cease some time before aggregation, usually as the result of the depletion of the food supply, and from that moment on the energy for the morphogenetic stages comes entirely from the reserves stored up in the vegetative stage. In the Myxomycetes the feeding also comes first and is followed by fruiting. This separation cannot readily be demonstrated in the Plasmodiophorales because of their parasitic habit, and thus far there is no evidence for it in the Labyrinthulales.

The fate of the cell mass following aggregation depends to a great extent upon the species. In some there is a period of migration of the cell mass of variable duration. Generally there are signs of differentiation of two cell types. The anterior cells begin to form a stalk at the apex. The stalk consists of large vacuolate cells enclosed in a delicate, tapering cylinder of cellulose. It is formed by the cells at the periphery of the apex moving up to the top and becoming trapped in the stalk proper. Once there, the amoebae begin to swell in a gradual conversion to the large pith-like stalk cells. As this occurs, the cellulose of the stalk is continuously deposited around them.

The posterior cells are to become the sorus. Again there is variation among species, but usually the whole posterior portion is lifted as one mass into the air, so that it forms an apical glob of cells when the stalk formation is completed. Each amoeba in this mass of cells becomes encapsulated in a hard cellulose spore case, ready for germination and the next generation.

These cellular slime molds represent an excellent example of aggregation organisms, for during the aggregation process there is a coming together of cells, and these cells may or may not be of precisely the same genetic constitution. It is in every sense a gathering, and not merely a concentration of protoplasm in one spot, the latter being more the situation in the Myxomycetes. Furthermore, as a result of the aggregation one organism is made out of many; in the span of a few hours, without the help of growth, the separate cells come together to form a unified, coordinated, multicellular individual.

Concerning the phylogeny of the Acrasiales, the old tradition that they bear some relation to the Myxomycetes should undoubtedly be abandoned. The principal differences between the two are the total absence of a flagellated stage and the absence of the syncytial plasmodium in the Acrasiales.

Another point of divergence is the clear-cut sexuality in the Myxomycetes, where the zygote nucleus gives rise to the whole plasmodium. There is still insufficient evidence to support the notion of sexuality in the Acrasiales, despite the work of Skupienski (1920) and Wilson (1952 et seq.), although the possibility that future work may show the cellular slime molds to be sexual cannot be neglected. However, if there is to be an analogy to the sexuality of Myxomycetes, syngamy should begin after spore germination, and the mass of vegetative amoebae which enter ultimately into aggregation should be diploid. The fact that there are usually seven chromosomes during vegetative division (a fact which I have been able to observe in a few cases) clearly indicates that the vegetative amoebae are haploid. It is my view that this difference is a major reason for considering the two groups to be totally independent in their origin.

What then is the origin of the Acrasiales? To the Myxomycetes we ascribed a flagellate ancestry; the same was true of the Plasmodiophorales, with the possibility that the specific flagellate was allied to the Chytrid fungi; the Labyrinthulales are also flagellated, and Smith's suggestion of their relation to the golden algae is reasonable. The most likely origin of the Acrasiales is from the free-living amoebae of the soil. There are many amoebae which have no flagellated stage and no established sexuality; their life cycle consists of a free-wandering stage followed by a period of encystment. This idea is by no means new, and a number of the earlier workers from de Bary[12] on have made this suggestion; it still has much to recommend it. It should be remembered that it is presumed that the free-living amoebae themselves stem from the flagellates, for there is a close relation between pseudopod and flagellum. This means then that all four groups of Mycetozoa

[12] *Comparative Morphology and Biology of Fungi, Mycetozoa, and Bacteria.* Oxford, Clarendon Press, 1887.

discussed thus far could have separate and independent origins from a flagellate ancestor. Obviously this feature is not unique with them, for different groups of algae, fungi, and other plants, as well as sponges and other animals, all have had similar independent origins.

6. Protosteliales

The fact that we are still uncertain about phylogenetic relations has been emphasized by the discovery of a number of new species by L. S. Olive (1965) and Olive and Stoiano-vitch (1966a). These organisms, which their discoverers have placed in a new order, differ from the Acrasiales in that they do not aggregate in any sense; yet their spore structure and in some cases their amoeba structure show definite similarities to some species of Acrasiales. This is especially true of the new genus *Protostelium* that will be described in detail in the next chapter.

While most Protosteliales have similarities to the other Acrasiales, the genus *Cavostelium* differs radically in that it has a flagellated stage. The third new genus in the order is *Schizoplasmodium*, which has sorocarps that resemble those in *Protostelium* and *Cavostelium* but is multinucleate at all times and has a definite plasmodium. These new species of Olive and Stoianovitch would seem in some ways to link the Myxomycetes and the Acrasiales, but perhaps we are not yet ready for any final Mycetozoan phylogeny.

7. Other aggregative organisms

It is a curious paradox that the organisms that most closely resemble the Acrasiales in their life history, the Myxobac-teria, are completely unrelated. Their similarity was recognized by Thaxter (1892), who discovered the group.

In the larger, more conspicuous members of the Myxobacteria such as *Chondromyces*, the resistant bodies are cysts which enclose many rod-shaped bacterial cells. If sown in a suitable environment, they split open and produce a stream of rods that soon begin their period of vegetative growth. During this phase there is a marked tendency for the cells to stick together and follow one another's tracks as they glide smoothly along. Therefore, in a sense, they aggregate continually, and if two groups are in close proximity they will merge. There is in this merger some evidence of mutual attraction which could conceivably be chemotaxis.[13]

Fruiting begins by a clump of cells becoming rounded, drawing itself up from the surface of the agar, and rising into the air. The bacterial rods produce much slime, which is left behind as the rods themselves seek an uppermost position, so that a stalk made up largely of this exudate forms in the rear. As the mass of rods advances it may bifurcate, ultimately producing a number of globular masses at the end of the branched, exuded stalk. Each one of the masses now becomes carved out by indentation into a cluster of cysts, and these cysts remain viable for long periods of time under unfavorable growth conditions.

If we search elsewhere for examples of aggregation organisms, we will find one case that is especially well known and well understood, namely the development of Ascomycetes. These fungi have the ability to produce hyphal fusions, and in this way there can be aggregations of nuclei of various genetic constitutions. The nuclei do not fuse but remain in the haploid condition, intermingling in the cytoplasm, a condition known as heterocaryosis. This fusion of genetic strains may not involve an immediate large concentration of proto-

[13] For experiments and references see J. T. Bonner, *Morphogenesis*, Princeton University Press (Athenaeum), (1952); A. McVittie and S. A. Zahler, *Nature 194* (1962): 1299-1300; W. Fluegel, *Proc. Minn. Acad. Sci. 30* (1963): 120-123; M. S. Quinlan and K. B. Raper, *Handbook of Plant Physiology*, Vol. XV/1, pp. 596-611. Springer-Verlag, 1965.

plasm, as was the case in the Acrasiales, but in a modified way this does occur. In the first place, it has been shown that the nuclei can move large distances fairly rapidly in the mycelial network; secondly, when spore formation occurs, either sexual or asexual, there is an aggregation or flow of protoplasm within the vegetative mycelium to the fruiting structure.

Other examples are harder to find. In a sense the development of the green alga *Enteromorpha minima* shows some aggregative properties. In related species of *Enteromorpha* the young plant develops from a single swarm cell that attaches to the ocean floor. In *E. minima* Bliding[14] showed that a whole mass of swarmers develops in an area, each producing young shoots, which in turn fuse to form one individual. It is a case in which there is no apparent inward migration of material or mutual attraction (at least none has been demonstrated among the swarmers), but the individual does have numerous parents and a polyglot genetic constitution.

8. The significance of aggregation in biology

The significance of aggregation has been discussed previously in some detail (Bonner, 1958), and here I would like to emphasize briefly three points.

The first is that the aggregation process may serve as a partial substitute for sexuality. As Haldane[15] has pointed out in the case of Ascomycetes, there is a gathering of diverse nuclei in the heterocaryon, and then by the production of many haploid spores these nuclei are segregated so they may again combine in other ways upon subsequent germination and fusions. Such a system has the advantage over sexuality that there can be more than two parents, but it lacks the advantage of chromosomal recombination that can result from

---

[14] *Botan. Notiser.* (1938) : 83-90.
[15] *New Biology 19* (1955) : 7-26.

the formation of diploid nuclei and the subsequent meiosis.

In the Acrasiales it is not merely a matter of nuclei, but of whole cells coming together in the aggregation. Yet the same point applies, since the nuclei of many parents are gathered in one sorus, which can then be redistributed in the uninucleate spores to form new combinations. The fact that the cellular slime molds have this substitution for sex, this recombination on a cellular level, does not mean that a truly sexual system might not exist as well; in fact, the Ascomycetes have both systems. On the other hand, it is conceivable that such a system of handling and recombining variants is adequate for the Acrasiales, as it is for many of the imperfect fungi. It should be added that all these arguments also apply to the Myxobacteria, the Labyrinthulales, and to the Myxomycetes in those cases where there is a fusion of plasmodia.

The second point concerning the significance of aggregation is the possibility that for some forms this is the channel by which a truly multicellular condition has been achieved. If one postulates that separate cells living in close proximity developed mutual deficiencies so that they became dependent upon each other, then it would follow that there would be a selective pressure in favor of an aggregative mechanism. There is, therefore, the possibility that cells with genetically determined nutritive deficiencies benefit from aggregation to some extent, in the way that cells benefit by sexual fusion. Aggregation as well as sexuality provide genetic advantages for association. This being the case, we have a rational understanding of a possible evolutionary reason for the existence of aggregation and the production of multicellular individuals by aggregation. Obviously there are other ways in which multicellular organisms came into being; aggregation is merely one of the possibilities.

The third and last point is the importance of aggregation as a tool in the study of the mechanisms of development.

There are many aspects of development which have remained refractory to interpretation, and of these perhaps the most important is the mechanism of regulation: the fact that a group of cells with equal potencies may be cut, fused, or redistributed, yet the cell mass as a whole will give rise to a perfect individual. One of the difficulties in attacking this problem has been that in conventional embryos which arise from a fertilized egg there is a mixture of growth and a segregation of potencies along with periods of partial regulation.

Aggregative organisms, and in particular the cellular slime molds, afford almost perfect examples of regulatory behavior. They are made up of groups of equipotential cells which do not arise in specific sites in an embryo, but grow separately and then come together by piling on top of one another. This suggests that the possibilities of cell arrangement during aggregation are numerous, yet each combination produces a normal individual. But, as will be shown, there is the possibility of cell rearrangement within the mass, and furthermore, despite the fact that the cells are equipotent, there is ample room for variation among them, and there is a good possibility that cell variation is involved in subtle ways in the process of differentiation. In fact, it will be important to examine the problem of variation on all levels and from all aspects in the Acrasiales in order to obtain a better understanding of both the mechanism of inheritance and the mechanism of development in these organisms.

II. The Cellular Slime Molds

1. A brief survey of the Acrasiales

THE FIRST comprehensive view of the cellular slime molds is the monograph on the Acrasieae by E. W. Olive, published in 1902. Since then some new species have been described, and a few of the original species have not been seen again since their first description. Some of the taxonomic and descriptive advances have been included in a number of papers by K. B. Raper (1940a, 1951, 1956a, 1960a) and in some recent papers by L. S. Olive (1962, 1965; Olive and Stoianovitch, 1960, 1966a,b,c,d). We still need a detailed, modern monograph on this group.

Here no attempt will be made to give proper taxonomic descriptions; rather I would like to give the information that will be of greatest use to the experimentalist and will list the important genera and species and then follow this list with a brief description of each.

Since, as has been emphasized, it is impossible to make any kind of certain phylogenetic scheme, it will be more helpful to emphasize structural affinities and dissimilarities. When this is done with the Acrasiales it is soon evident that it is very doubtful that even all members of this group arose from a common stock; they are most likely polyphyletic. This has been appreciated from E. W. Olive (1902) onward and has again been emphasized by Raper (1960a) and L. S. Olive 1965; Olive and Stoianovitch, 1960).

The principal basis for such speculation is that the amoebae of different types of cellular slime molds are radically different in appearance. *Guttulina* and *Guttulinopsis* have limax-type amoebae with blunt, lobose pseudopods, while the amoebae of *Dictyostelium* and their relatives have numerous, slender, elongate pseudopodia. Furthermore, *Sappinia*, also included in the Acrasiales because there is some cell aggregation, has large amoebae that differ significantly in that they are consistently binucleate. Since we have included *Sappinia,* it might also be reasonable, following Raper's (1960a) example, at least to mention *Hartmanella,* which has been described in detail by Ray and Hayes (1954), for it also has a primitive "aggregation" of amoebae, again differing in appearance from the others in that they are large and uninucleate, and have many small filose pseudopodia. However, if aggregation is the criterion for the inclusion of an organism in the Acrasiales, the new single-celled genus of L. S. Olive (1962), *Protostelium,* would have to be separated. There are other grounds for separating it: Olive has shown that its nuclei differ from those of *Dictyostelium.* On the other hand, some systematists might throw out *Sappinia,* and certainly *Hartmanella,* and consider that the Acrasiales consist of two families, possibly unrelated, the Guttulinaceae and the Dictyosteliaceae. In the scheme presented here I shall arbitrarily include all those amoeboid organisms that lack any flagellated stage and that show some form of aggregation, however simple, and furthermore I shall include with them the non-aggregating forms that distinctly resemble them and are presumed to be close relatives. The major subdivisions will be based simply on the construction of the amoeba itself, the form groupings being:

  i. The *Hartmanella* amoeba: large with many filose pseudopods
 ii. The *Sappinia* amoeba: large binucleate, with broad, lobose pseudopods
iii. The *Guttulina* amoeba: small limax-type amoebae
 iv. The *Dictyostelium* amoeba: small amoeba with a number of filose pseudopods

Let me emphasize that this is not presented as a system of taxonomy, but only as a provisional convenience for the

experimental worker. When we know more of the aggregation mechanism in these groups, especially in *Hartmanella*
and *Sappinia,* we may find new reasons for including or ex-
cluding them from the Acrasiales. Let us now list all the
genera and some important species (Fig. 2):

i. The *Hartmanella* amoeba
   Hartmanella. A genus of common soil amoebae in
      which simple aggregation is described in one
      species, *H. astronyxis,* by Ray and Hayes
      (1954).
ii. The *Sappinia* amoeba
   Sappinia. Dangeard (1896). Common.
iii. The *Guttulina* amoeba
   Guttulinopsis. E. W. Olive (1901). Olive describes
      three species in this genus, and recently the
      genus has been re-examined by Raper (1960)
      and L. S. Olive (1965).
   Guttulina. Cienkowsky (1873). Four species in this
      genus were described by Cienkowsky, Fayod
      (1883), and van Tieghem (1880), but the
      first detailed description is that of Raper
      (1960a) of a new species, and further descrip-
      tions should be forthcoming.
   Acrasis. van Tieghem (1880). This had not been seen
      since the first description until the recent
      work of Olive and Stoianovitch (1960), who
      described a new, second species.
iv. The *Dictyostelium* amoeba
   Protostelium. L. S. Olive and Stoianovitch (1960).
      Three species have been described (L. S.
      Olive, 1962). Since the nuclei differ from
      those of *Dictyostelium,* there is a possibility
      that this should constitute a separate, fifth
      type of amoeba.

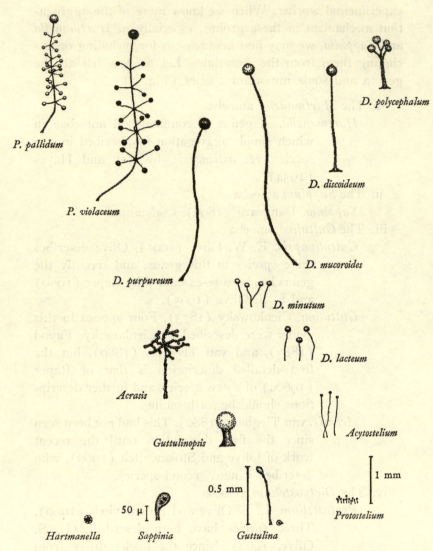

Fig. 2. The mature fruiting bodies of different species of Acrasiales (and related forms).

Acytostelium.

    Acytostelium leptosomum Raper (1956a). Rare.

Dictyostelium

    Dictyostelium lacteum van Tieghem (1880). Unknown until it was rediscovered by Raper (1951). Rare.

    Dictyostelium minutum Raper (1941a). There is the possibility, suggested by Raper (1941a), that this species is a diminutive *D. mucoroides* and therefore should appropriately be part of the mucoroides "complex." Relatively infrequent.

    Dictyostelium mucoroides Brefeld (1869). This is perhaps best described as a "complex," for there is great variation in this group (Raper, 1951). Probably a number of the *Dictyostelium* species listed by E. W. Olive fall within this complex, as may also *D. giganteum* of Singh (1947a). Very common.

    Dictyostelium purpureum Olive (1901). Common.

    Dictyostelium discoideum Raper (1935). Isolated relatively infrequently.

    Dictyostelium polycephalum Raper (1956b). Rare.

Polysphondylium

    Polysphondylium violaceum Brefeld (1884). Very common.

    Polysphondylium pallidum Olive (1901). Olive also lists another white-spored species of *Polysphondylium*, but, as Raper (1951) points out, this quite likely merely represents variation in the *pallidum* group. Common.

There is one genus that is known only from its original description and unfortunately has not been rediscovered:

    Coenonia. van Tieghem (1884).

This form has a peculiar cup-like structure holding the spore mass, and at the base of the stalk there is a branched hold-fast. Recently Raper (1960a) has discovered a new species of *Dictyostelium* (still formally undescribed and unnamed), which has a normal *Dictyostelium* sorus and a holdfast similar to van Tieghem's description of *Coenonia* at the base of the stalk.

2. *Hartmanella astronyxis*

These large soil amoebae will aggregate into definite groups before encystment, as was shown by Ray and Hayes (1954). Nothing is known of the mechanism of this clumping, but it is a striking and clearly defined phenomenon. All the cells in the clumps form cysts or spores. There is no suggestion of a stalk in these forms and all the cysts lie side by side in a flat disc. The new cycle begins when the individual cysts split open and a uninucleate amoeba emerges from each.

3. *Sappinia*

Sappinia has two characteristics which give it its affinity to the Acrasiales. In the first place it forms aggregations of a crude sort, somewhat similar to *Hartmanella*. Secondly individual cells, under some conditions, are raised upon stalks so that the cell and its stalk together have a club-like appearance.

The aggregations are poorly understood. Raper (1960a) briefly mentions some interesting new observations. It has been noted previously (E. W. Olive, 1902) that the aggregations tend to appear upon ends of sticks or bits of debris, and that the cells in the aggregate all become rounded and are encysted in a firm wall. Raper points out that an aggregation can occur during feeding and that the amoebae sometimes dissociate before spore formation. Also the aggregate may move as a hemispherical unit over the surface of the substratum. No work has been done on the mechanism of aggregation.

The stalked resting stage of the individual cells is different from any of the stalked structures in other Acrasiales. This so-called stalk is really no more than a narrowed portion of the protoplast, and furthermore it has not been possible to demonstrate any persistent material there. Raper (1960a) reports that he has found such structures in a variety of soil amoebae, some of which are clearly not *Sappinia*, being uninucleate rather than binucleate, and makes the point that this emphasizes the need for a great deal of work on the taxonomy of these "lower" or borderline Acrasiales as well as of soil amoebae in general, before the problem of what to include in the Acrasiales, and what to exclude, can be properly settled.

The point of greatest difference between *Sappinia* and the other members of the Acrasiales is the fact that *Sappinia* is binucleate. This is interesting because of the possibility that here may be a case of a true heterocaryon (or dicaryon) in an amoeba. *Sappinia diploidea* has been studied by Hartmann and Nägler, and according to Wenrich it represents the only acceptable case of demonstrated true sexuality in free-living amoebae, although even here there remain some points that need further clarification.[1] The haploid nuclei of *Sappinia* lie side by side and divide simultaneously at each binary fission. During the sexual process two amoebae come together to form a common cyst. The two nuclei of each amoeba then fuse (as in the delayed caryogamy of Basidiomycetes), and then finally the protoplasts fuse. Both diploid nuclei now undergo two reduction divisions, during each of which one of the daughter nuclei degenerates. The final two remaining haploid nuclei are the normal complement of the vegetative, dicaryon individual which then emerges from the cyst.

[1] For references see D. H. Wenrich, pp. 134-265 in *Sex in Microorganisms*. American Association for the Advancement of Science, Washington, D.C., 1954.

Although interesting in its own right, there is nothing in this cycle that has a counterpart in other members of the Acrasiales. Even the cytological observations of Wilson (1952, 1953; Wilson and Ross, 1957) on *Dictyostelium discoideum* could in no way be homologized with this work on *Sappinia*. It is true that occasionally in stained preparations of higher cellular slime molds it is possible to see binucleate cells, but these are very infrequent and must be presumed to have occurred accidentally and to play no role in the normal life history. Furthermore, in such cases the two nuclei do not lie side by side as in *Sappinia*, but are separate in the common protoplasm.

4. *Guttulinopsis* and *Guttulina*

The only modern treatments of these two genera are those of Raper (1960a) and L. S. Olive (1965), and it is evident, as these two authors point out, that we still know very little about the relationships of the genera to each other. The reason they are grouped together here is that there remains some doubt as to how they differ and we must await a detailed analysis.

Long ago E. W. Olive (1902) suggested that the principal distinction between the two genera was that *Guttulinopsis* did not have true spores, but crinkled, hardened amoebae, which he called pseudospores. In a recent study of the common *Guttulinopsis vulgaris*, L. S. Olive (1965) suggests that this is not a valid distinction and that the misshapen, resistant cells are true spores. He considers more significant distinctions the fact that the cells in the sorocarp of *Guttulinopsis* are grouped in compartments and the fact that many of the cells within the stalk compartments appear dead and degenerating. This leads him to quite contrary identifications of specific forms from Raper (1960a). These differences will surely be resolved, but for our present purposes it may be best simply to lump them all (including the new forms of Raper and Olive) into a "*Guttulina* complex" and await the further studies of these investigators.

In all these *Guttulina*-type organisms, aggregation consists of a slow grouping together of cells; there is often no inflowing of streams. In all cases this results in a considerable movement of cells, but the actual fruiting process is varied and poorly understood. Raper (1960a) describes a species that pushes up a filament of cells (more than one cell thick) that often has a slight knob at the tip. The cells appear loosely organized, and if two fruiting buds bump into one another they will fuse. In this particular form there are no predetermined spore and stalk cells in definite regions; all the cells apparently serve both functions (Fig. 3).

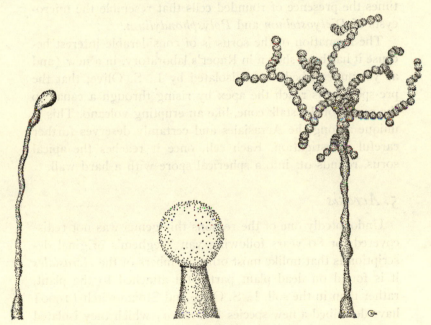

Fig. 3. *Guttulina, Guttulinopsis,* and *Acrasis* (from left to right). See p. 26 for problems concerning identity of *Guttulina* and *Guttulinopsis*. (After L. S. Olive and K. B. Raper.)

THE CELLULAR SLIME MOLDS

In another form described in detail by L. S. Olive (1965) there is a moderately clear separation of the spore and stalk regions. The spores are thin walled, irregular, with flattened or concave sides and are found primarily within compartments in the apical region. The compartments in the stalk are irregular, containing some spores and numerous degenerating cells.

In one interesting form described by Raper (1960a) there is a short conical stalk surmounted by a spherical sorus (Fig. 3). The mechanism of the formation of the stalk is obscure, as is its ultimate cellular fate. There is no evidence of rigid stalk cell walls as found in *Dictyostelium*, but instead an inconsistent picture indicating the retention of undifferentiated amoebae, the degeneration of some cells, and sometimes the presence of rounded cells that resemble the microcysts of *Dictyostelium* and *Polysphondylium*.

The formation of the sorus is of considerable interest because it has been shown in Raper's laboratory, in a new (and as yet undescribed) form isolated by L. S. Olive, that the pre-spore cells reach the apex by rising through a canal up the center of the stalk cone, like an erupting volcano. This is unique among the Acrasiales and certainly deserves further careful investigation. Each cell, once it reaches the apical sorus, rounds off into a spherical spore with a hard wall.

5. *Acrasis*

Undoubtedly one of the reasons this genus was not rediscovered for 80 years following van Tieghem's original description is that unlike most other members of the *Acrasiales* it is found on dead plant parts still attached to the plant, rather than in the soil. L. S. Olive and Stoianovitch (1960) have described a new species (*A. rosea*) which they isolated from the dry florets of a grass, and it is now possible to gain an accurate picture of *Acrasis*. The amoebae of this form resemble *Guttulinopsis* and *Guttulina* rather than *Dictyos-telium,* but because of the branching sorocarp and distinct stalk and sorus, they suggest that this genus is related to *Guttulina* and its associates, but is more complex (Fig. 3).

The aggregation is much the same as for other members of the Guttulinaceae: a streamless accumulation of many cells into a lump. Following aggregation stalk formation begins. The stalk appears to be deposited apically as a bleb of cells rises with the tip of the forming stalk. Ultimately the apical bleb of cells thins out into a series of branching chains of spores so that the overall structure has a tree-like appearance; the trunk is the stalk and the branches the uniserate chains of spores. The stalk cells undergo little differentiation and somewhat resemble the spores, although the latter have ring-like thickenings at the points of attachment between cells. In *A. granulata* of van Tieghem it is clear that the spores are more highly differentiated, showing pigmentation of the wall and an irregular surface.

Some most interesting observations have been made on an ephemeral type of recombination in *A. rosea* by L. S. Olive, Dutta, and Stoianovitch (1961). They isolated a number of strains which possessed different capacities to ingest different species of yeast and other microorganisms, and then grew two strains with maximum divergence in food utilization on a form that they can both eat. The spores from this generation were tested upon the whole spectrum of food organisms, and 3 to 4 per cent were found to have undergone some sort of recombination, for they now possessed the dietary capabilities of both parents. However the authors point out that, since this recombined condition only lasted five or so generations, after which the amoebae reverted to one of the two parental types, this cannot be considered evidence for sexuality. They suggest instead, as one of a number of possible explanations, that the cells show partial, temporary fusion in anastomosis, which they have demonstrated to occur, and that there is some cytoplasmic exchange as a result of this anastomosis, which becomes diluted out after a few generations.

6. *Protostelium*

In this new genus of Olive and Stoianovitch (1960) there is no aggregation whatsoever. The individual amoebae, which closely resemble those of *Dictyostelium,* may either form rounded cellulose cysts right on the surface of the substratum, similar to the microcysts of *Dictyostelium* and *Polysphondylium,* or they may form remarkable stalked structures. A single amoeba will become hemispherical and secrete a minute slender stalk, and in the process the amoeba is raised into the air 70 μ at the most (Fig. 4). In *Protostelium mycophaga,* the cell at the apex becomes rounded and has a very thin wall. If the stalk bends and this spore touches the agar, or if the spore falls off, it soon resumes the amoeboid shape, sometimes leaving behind a thin spore wall. In *P. arachispora* the spore is elongate and it can reabsorb the stalk thereby lowering itself back to the substratum to renew a vegetative existence.

As Olive and Stoianovitch (1960) point out, there are some reasons to include this genus among the Acrasiales, although there are other good reasons for making it into a separate order, the Protosteliales. Besides some resemblance of the amoebae to those of *Acytostelium, Dictyostelium,* and *Polysphondylium,* the microcysts have a cellulose wall, as do those of these three genera. Finally, the stalk closely resembles that of *Acytostelium* (which we shall examine next) in that they are both non-cellular; also in some stalks of *Protostelium* there is a slender lumen near the base which resembles that of *Acytostelium.* B. M. Shaffer has informed me that he has observed in sparse cultures of *Acytostelium,* occasional fruiting structures made up of a single cell, almost exactly like those of *Protostelium*.

7. *Acytostelium leptosomum*

This new species and genus of K. B. Raper (1956a; Raper and Quinlan, 1958) is of great interest, because it is the only

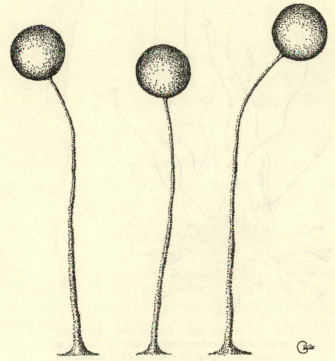

Fig. 4. *Protostelium.* Fruiting bodies of *P. my-cophaga.* A single spore is held in the air by a thin stalk about 70μ high. (After L. S. Olive.)

aggregating cellular slime mold that produces a totally non-cellular stalk. The stalk is a very thin tube of cellulose bearing one round spore mass at its tip (Fig. 5).

The spores are spherical as in *Dictyostelium lacteum,* and

they germinate by a dissolution of part of the surface and a swelling of the protoplast. Once emerged, the amoebae resemble those of *Dictyostelium*, but as is true of other small

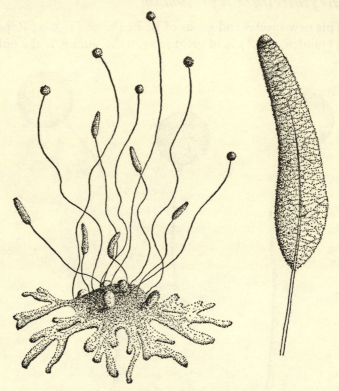

Fig. 5. *Acytostelium leptosomum. Left,* the simultaneous occurrence of aggregation and culmination; *right,* an enlarged culminating pseudoplasmodium showing the acellular stalk and the cell orientation. (Drawings based primarily on photographs by Raper and Quinlan, 1958.)

and delicate species, they require a weak culture medium and are surprisingly rigid in their optimal culture requirements. Even when growth and development is at its best, a large number of microcysts are normally produced, which gives rise to an interesting source of confusion. Both the microcysts and the spores are spherical and surrounded by a cellulose wall, and, according to Raper and Quinlan, the size range overlaps considerably (the microcysts being slightly larger on the average); therefore it is impossible to decide whether any one rounded body is a true spore or a microcyst (Fig. 7).

Aggregation on the whole is typical of Acrasiales, although some features are peculiar to *Acytostelium*. In particular, large networks of streams form which then fail to develop into sorocarps but eventually either break up into microcysts or "disintegrate," to use Shaffer's term. Smaller aggregation patterns, which form in the usual fashion, rarely give rise to one sorocarp, but usually up to twenty-five or more. Each fruiting body begins in the form of a small papilla; as culmination proceeds, aggregation continues for a considerable period (Fig. 5). Raper and Quinlan point out that the overlap of aggregation and culmination, as well as the production of numerous sorocarps in one aggregation pattern, are features found to a lesser degree in other small forms, i.e. *Dictyostelium lacteum* and *D. minutum*.

Each papilla rises into the air leaving a very thin hair of a stalk behind, and the cell mass has the shape of a spindle. The cells within the mass are oriented almost entirely at right angles to the stalk, and at the anterior end there appear to be one or two "cap cells," as Raper calls them, which cover the tip completely, like keystones. Down the central axis of the cell mass there is a very fine hole, and if looked at in cross section the cells appear wedge-shaped, coming to a point at this central hole.

The hole, of course, is the site of deposition of the stalk, which passes out posteriorly. The stalk is only 1 to 2 μ in diameter, and its length is from about 750 to 1500 μ. Raper and Quinlan have shown that it is a tube of cellulose with a minute lumen.

With some material that was kindly sent to us by Raper, we have made periodic acid–Schiff preparations for non-starch polysaccharides and find that there are few signs of differentiation or specialization within the cells; they take up the stain uniformly. Ultimately, after the stalk has reached its full height, all the cells become rounded spores. There is here no division of labor among the cells; all the cells produce the stalk and then all the cells produce the sorus.

8. *Dictyostelium lacteum*

This species is small and differs from other species of the genus *Dictyostelium* in that the spores are rounded instead of elliptical, closely resembling the spores of *Acytostelium*. The aggregation pattern is simple and shows definite streams. As soon as aggregation is completed the rounded mass of cells, unless the aggregates are very small, breaks up into a series of small blebs each one of which will produce a slender fruiting body.

In this case we have for the first time a true cellular stalk that is characteristic of all the species of *Dictyostelium* and *Polysphondylium*. The bleb becomes elongate and the anterior cells start forming central, large, vacuolate stalk cells. The pre-stalk cells are added to this initial group by moving up the sides of the central stalk cells to the apex. As the pre-stalk cells go up the side they secrete the cellulose stalk sheath and then enter into the top of the sheath they have constructed. As soon as they are inside they begin to swell and become vacuolate and secrete a further hard cell wall before they die.

The stalk is usually delicately tapered and in small forms such as *D. lacteum* (or small individuals of larger forms) the cells at the tip of the stalk may be in a single row, while toward the base the stalk may be more than one cell thick. The interesting fact is that the taper exists even when the whole stalk is made of single cells: one cell may, in a very narrow tip of a stalk, span a region of some length, while further down the cells will be short flat cylinders (or, more accurately, truncated cones) (Fig. 22).

In a study involving very small fruiting bodies it was noticed, in *D. lacteum*, that sometimes there would be an abrupt size reduction in the minute stalk. When these stalks were examined under the microscope it was clear that the sudden constriction was a junction between a lower cellular and an upper acellular stalk (Bonner and Dodd, 1962a). In fact the upper end of the especially small sorocarps resembled closely those of *Acytostelium,* especially since both had rounded spores, a fact which suggests an affinity between these two forms.

9. *Dictyostelium minutum*

A number of workers have pointed out that *D. mucoroides* includes many variants and may properly be described as a "complex." Raper (1941a) has suggested that *D. minutum* is part of such a complex, but because this small form appears so frequently in nature and because it has such special characteristics of its own (perhaps related to its size) it is convenient to consider this a separate species.

The aggregation process of *D. minutum* has recently been described in some detail by Gerisch (1963, 1964b) (Fig. 21). Undoubtedly the culture conditions and the particular strains used greatly affect the appearance of the aggregation pattern. In Gerisch's experiments the first centers of *D. minutum* began as streamless accumulations of cells, reminiscent of *Guttulina* and other primitive forms. Of particular interest is his observation that each accumulation is clearly around one central keystone cell, similar in appearance to the founder cells of Shaffer (1961b) in *Polysphondylium violaceum*. In some cases, as we have described for *D. lacteum,* this center breaks up into a series of small fruiting bodies, each one of which resembles those of *D. lacteum* except for the fact that the spores are elliptical or capsule-shaped (Fig. 21). In other cases the original center disintegrates and the cells move out radially. As they move out they begin to form streams, and ultimately new secondary centers are formed. Finally, in some cases, both the primary centers and the peripheral secondary centers may share the amoebae which move about in streams and flow to the various centers.

The rest of the development, once the blebs have formed, is similar to that in *D. lacteum.* The small fruiting bodies go straight up into the air and never travel along the surface of the agar. The fact that all the stalks are erect is apparently associated with their size, for if the normally larger forms such as *D. mucoroides* are made to form small fruiting bodies by reducing the amoeba density they also are erect and in fact indistinguishable from *D. minutum.*

10. *Dictyostelium mucoroides*

The process of aggregation and other characteristics of the life cycles of the cellular slime molds were studied in great detail by many of the earlier workers, who confined their efforts almost entirely to *D. mucoroides.* This is true of the earlier observations of Brefeld (1869, 1884) and van Tieghem (1880). At the turn of the century E. W. Olive (1902) contributed to a further clarification of the life cycle of this species, but the most remarkable experimental study is that of Potts (1902), who did much to gain insight into the physiological conditions of the development process. In more recent years a number of authors have added further descriptive details concerning the development of *D. mucoroides* (e.g. Harper, 1926; Arndt, 1937; Shaffer, 1956 et seq.; Bonner and Dodd, 1962a).

In general the larger forms of *Dictyostelium* and both species of *Polysphondylium* have similar aggregation patterns, although at the same time each form has certain detailed characteristics that separate it from the others. Furthermore, as mentioned earlier, the culture conditions, especially those affecting amoeba density and degree of moisture, will greatly influence the appearance of the aggregation pattern. I will not go into detail here, nor can any reference be given that contains a comprehensive account of all these facts, although characteristic differences are pointed out and often illustrated by most authors.

There is as yet no evidence for founder cells in *D. mucoroides; D. minutum* is the only species of *Dictyostelium* which appears to have them (Gerisch, 1946b, 1965h). Aggregation is characterized by long streams that may sometimes extend considerable distances. If the conditions are especially moist, or if the amoebae are under a thin layer of water, there may be centerless orientation of the amoebae over wide areas, a condition first described by Arndt (1937) and referred to by him as the "preplasmodium" stage. Ultimately centers are formed and the amoebae reorient toward these in ever thickening streams.

One striking feature in *D. mucoroides*, and all the other species that follow, is that in the transition from the feeding or vegetative stage to the aggregation stage there is a marked change in the shape of the amoebae; they go from a relatively flat, isodiametric shape to an elongate spindle shape. In *D. minutum* this change takes place after the foundation of the primary centers, while in *D. mucoroides* it usually takes place in the preplasmodium stage, before any centers appear.

During the latter part of aggregation a cone rises above the center, and immediately the stalk is formed at the apex.

As the tip of the stalk is continually added to, the whole cell mass becomes sausage-shaped and moves along the surface of the substratum, showing marked orientation toward light. At this point aggregation may not be completed and the last aggregation stream will flow along the stalk into the cell mass. In conditions of excessive moisture (which will be discussed in detail later) the migration may continue for many hours along the agar surface, and during this time the pre-stalk cells are clearly distinguishable, in histological sections, from the posterior pre-spore cells. Eventually the tip will turn upward and rise into the air, and the last of the pre-stalk cells will turn into stalk, while all the pre-spore cells will become encapsulated in their elliptical cellulose cylinders (Fig. 6).

It may be helpful to pause here and mention some of the various terminology found in the literature. The aggregate of amoebae has been called the "cell mass," or the "pseudoplasmodium," and more recently Shaffer (1962) has introduced the term "grex" to describe those stages where there is a definite shape to the cell mass, with an axis of symmetry. In *D. discoideum* there is a period of stalkless migration of the cell mass and this stage has been called the "migrating pseudoplasmodium," the "migrating slug," or the "migrating grex." In *D. mucoroides* and other species that have a stalk from the beginning there is some disagreement as to whether the period during which the stalked cell mass is crawling over the surface of the substratum should be considered a period of migration or a period of culmination. My prejudice is in favor of considering it migration, but Shaffer has attempted to circumvent this issue by calling one stage the "lying grex" and the other the "standing grex." As long as one's meaning is clear, the particular choice of words is relatively unimportant. It should be added that sometimes the term "sorogen," devised by Harper (1926) to indicate

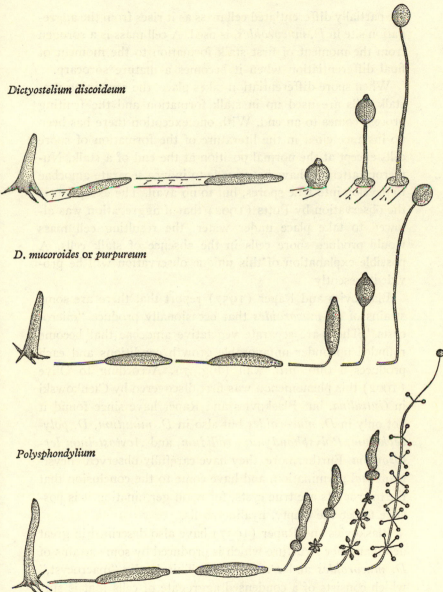

Dictyostelium discoideum

D. mucoroides or *purpureum*

Polysphondylium

Fig. 6. The migration and culmination stages of three different types of cellular slime molds. (From Bonner, 1958.)

the partially differentiated cell mass as it rises from the aggregation site in *D. mucoroides*, is used. A cell mass is a sorogen from the moment of first stalk formation to the moment of final differentiation when it becomes a mature sorocarp.

When spore differentiation takes place, the remaining prestalk cells are used up in stalk formation and the fruiting process comes to an end. With one exception there has been no instance cited in the literature of the formation of spore cells except at the normal position at the end of a stalk. Numerous attempts have been made to induce separate amoebae to develop into true spores, but to no avail. The exception is the observation by Potts (1902) that if aggregation was allowed to take place under water, the resulting cell mass would produce spore cells in the absence of stalk cells. A possible explanation of this unique observation will be provided presently.

Blaskovics and Raper (1957) report that there are some strains of *D. mucoroides* that occasionally produce "microcysts." These are separate vegetative amoebae that become rounded up under unfavorable growth conditions and each produces a thin, firm wall (Fig. 7). According to Olive (1902) this phenomenon was first discovered by Cienkowski in *Guttulina*, but Blaskovics and Raper have since found it not only in *D. mucoroides* but also in *D. minutum, D. polycephalum, Polysphondylium pallidum,* and *Acytostelium leptosomum*. Furthermore, they have carefully observed encystment and germination, and have come to the conclusion that the microcysts are true cysts, for upon germination it is possible to see the empty, hyaline shells.

Blaskovics and Raper (1957) have also described in great detail another structure which is produced by some strains of *D. mucoroides* and *D. minutum*. This is the "macrocyst," which consists of a condensed aggregate of cells that is surrounded by a thickened wall (Fig. 7). This was first observed by Raper a number of years ago, and he suggests that the "dwarfed sporangia" of Brefeld (1869) were most likely macrocysts.

In a careful examination of the role of macrocysts in the life cycle, Blaskovics and Raper (1957) come to the following conclusions. They form as the result of a small but normal aggregation; the aggregated mass, instead of heaping up and forming a stalk, rounds up into irregular spheres of cell masses, and each sphere becomes covered with a heavy cellulose wall. After a period the boundaries of the compressed, polyhedral cells within the macrocyst are no longer visible; this stage is followed by a contraction of the internal protoplasmic material. Blaskovics and Raper did not ascertain the nuclear and cellular nature of the protoplasm during these stages, but the evidence is excellent, pending histological and cytological studies, that these macrocysts do germinate and produce myxamoebae. The environmental conditions, especially the temperature and the composition of the medium, were found to affect differentially the formation of normal fruiting bodies and spores as opposed to macrocysts. Even the conditions which favor germination of spores and macrocysts appear to differ, depending to some extent on the strain. They also were able to demonstrate that macrocysts are more resistant to adverse environmental conditions than myxamoebae, but not so resistant as true spores. On the basis of these facts Blaskovics and Raper suggest that possibly macrocysts represent an alternative resting stage; that is, in the species *D. mucoroides* there may be three kinds of resistant bodies: microcysts, macrocysts, and true spores.

It is possible to show that macrocysts can form under water, and their function may therefore be an alternative resting stage in aquatic conditions. Since normal spores cannot form under water, yet aggregation does occur under these circumstances, it would indeed be advantageous for the organism to produce resistant bodies; in seasons of high rainfall a suitable dry surface for normal development might not exist. When a strain forming macrocysts was immersed in water, beautiful clusters of macrocysts were produced in the

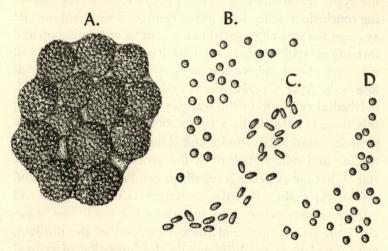

Fig. 7. Different types of resistant bodies of the Acrasiales. A, macrocysts; B, microcysts; and C, spores of *D. mucoroides*; D, spores of *D. lacteum*.

bottom of the dish. Here we have not only a possible explanation for the function of macrocysts but also an indication that the underwater spores of Potts (1902) might have been macrocysts. Also related is the new observation of Weinkauff and Filosa[2] that the macrocysts are occasionally stalked.

It should be mentioned that Blaskovics and Raper showed that some strains that formed macrocysts could, by single-cell isolation, be separated into clones which differed in their ability to produce macrocysts. The only other species that produces macrocysts, according to these same authors, is *D. minutum*, again showing its affinity to *D. mucoroides*.

[2] Personal communication.

11. *Dictyostelium purpureum*

Superficially *D. purpureum* closely resembles *D. mucoroides* except for the fact that in their final differentiation the spores turn purple. The intensity of the color varies in different isolates, and in some cases the spore heads are almost black. There is no evidence for any great size variation in *D. purpureum* as there is in the *D. mucoroides* complex, nor are there any known strains that produce macrocysts.

There is one interesting phenomenon common to both *D. mucoroides* and *D. purpureum,* but it was first discovered in the latter species by Raper and Thom (1941). They showed that certain strains periodically have interruptions in their stalk formation as the cell masses migrate over the surface, leaving a series of gaps in the stalk. These may be exaggerated under certain cultural conditions (and even produced in strains where it had not previously been observed). In general, those conditions which favor migration favor the interruption of the stalk, but the correlation is not perfect, and further work is needed to settle the point. Its importance here is that it suggests the relation between *D. mucoroides* and *D. purpureum*, which have a period of migration with stalk, and *D. discoideum*, which utterly lacks it.

12. *Dictyostelium discoideum*

More is known by far of the morphological details of this species than of any other (Figs. 6, 8; Plates 5, 6, 7, 8). Ever since its discovery by Raper in 1935, its potentialities as an experimental organism have been appreciated and it has thus been intensively studied.

A host of different methods have been used in the descriptive work. The groundwork was laid by Raper (1935, 1940a, 1940b, 1941b; Raper and Fennell, 1952), who, by culturing the organism on different media and under different environ-

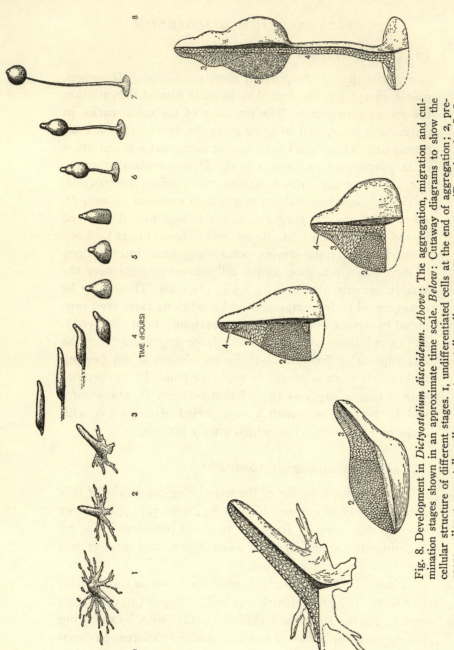

Fig. 8. Development in *Dictyostelium discoideum*. *Above:* The aggregation, migration and culmination stages shown in an approximate time scale. *Below:* Cutaway diagrams to show the cellular structure of different stages. 1, undifferentiated cells at the end of aggregation; 2, pre-spore cells; 3, pre-stalk cells; 4, mature stalk cells; 5, mature spores. (Drawing by J. L. Howard, courtesy of the *Scientific American*.)

mental conditions, by making stained whole mounts, and by cutting and grafting experiments, was able to obtain a comprehensive picture of the life cycle, to which his motion picture of the migration phase added some details.

Following the lead of Arndt (1937) with his motion picture of *D. mucoroides* (taken in 1929), I made in 1941 a time-lapse film of the whole cycle of *D. discoideum* (and have added to it subsequently). This proved most helpful in understanding many of the details of development.

The use of histological sections has also been rewarding (Bonner, 1944), for in this way it is possible to follow with some accuracy the internal cellular details. At first, rather standard methods using haemotoxylin were employed, but later more modern histochemical methods were found most helpful (Bonner, Chiquoine, and Kolderie, 1955). This approach also has been used by Krivanek (1956) along with some biochemical analyses.

There has also been some work at the cytological level, especially on the nuclear phenomena (Wilson, 1952, 1953; Wilson and Ross, 1957; Bonner and Frascella, 1952; La-budde, 1956), and finally, there are some useful studies of *D. discoideum* with electron micrographs of ultrathin sections and other preparations by Mühlethaler (1956), Gezelius and Rånby (1957), Mercer and Shaffer (1960), and Gezelius (1961).

The following description of each stage of the life cycle of *D. discoideum* is based primarily on the references given above. I will not attempt to specify everywhere the author responsible for the original observation; my object is to describe the slime mold rather than write a history. Also it should be mentioned that occasional observations were made originally on some other species and the information transferred to *D. discoideum*. This is an obvious consequence of the fact that *D. discoideum* has been discovered relatively recently.

Spores and spore germination in *D. discoideum*

The spores of *D. discoideum* are elliptical in shape. They vary considerably in size, being approximately 6 to 9 μ long and 2.5 to 3.5 μ in diameter. The nuclear size, which may be easily measured in Feulgen preparations, shows a range of diameters from 0.8 to 2.5 μ. Nuclear stains reveal little in the way of chromosome structure at this stage.

When the spores are stained with toluidine blue, a tightly knit group of metachromatic granules can be seen lodged alongside the nucleus. The significance of this striking staining reaction is unfortunately unknown, and we can only surmise that the granules are polysaccharide particles of unknown function. In the other stages of the life cycle, except the stage of active feeding where metachromasia is totally absent, the granules are randomly distributed through the cytoplasm rather than in one closely integrated bunch.

With his electron micrographs Mühlethaler (1956) has been able to demonstrate clearly that the spore wall is made up of two layers, a double membrane. He believes that the outer one is condensed slime material of unknown chemical nature (similar to the slime sheath), while the inside wall is an intercrossed lacework of cellulose micelles which presumably give the spore its rigid framework. He points out that such a double membrane is common among microorganisms, but as yet its significance is not fully appreciated.

A few hours after the spore is sown in a favorable environment there is a split down the side of the capsule and the amoeba emerges. The process can be seen with considerable clarity in Arndt's motion picture. He used *D. mucoroides,* but the process of germination is the same in *D. discoideum.*

The vegetative stage of *D. discoideum*

Another unfortunate gap in our knowledge is in the nuclear behavior of the vegetative amoebae. As mentioned previously, I have observed a few vegetative divisions and was able to find the haploid number of seven chromosomes. The difficulty is that the bacterial food supply picks up the nuclear stains avidly, and many hundreds of amoebae have to be examined critically before it is possible to find one that is, with any certainty, in the process of division. The matter is especially tantalizing because of the curious report of E. W. Olive (1902) that the first nuclear divisions following spore germination in *D. mucoroides* are rather special and perhaps different from the subsequent divisions. This, of course, is quite possibly erroneous, but it should be checked along with many other details of the nuclear behavior.

Other cytological studies of vegetative amoebae have not revealed any very startling facts, with the possible exception of the absence of metachromasia mentioned previously. This presumably has some connection with the feeding process, but the nature of the connection is unknown. Another point of interest that has turned up in some preparations is that it is not uncommon to find evidences of cannibalism, a phenomenon well known among amoebae in general. In some instances it is possible to see more than one amoeba inside another; in one case I have observed three amoebae, clearly within food vacuoles, inside one large cannibal. This matter has been examined in detail by Huffman and L. S. Olive (1964). The observation is relevant to those of Wilson (1952, 1953; Wilson and Ross, 1957), who interprets one amoeba surrounding another as fertilization. That the cases mentioned above were cannibalism and not syngamy is borne out by two facts: (1) more than one amoeba was inside the other (although this could be argued to be polyspermy), and (2) the nuclei of the ingested amoebae are shown in various stages of digestion. One case has been observed of an amoeba in an aggregation stream with another partially digested amoeba inside it.

Another cytological detail worth mentioning is that, under a phase microscope and with Bodian's silver impregnation method, a polar cap on the nucleus is evident. In some instances in the silver impregnation slides it is possible to see a strand extending out into the cytoplasm. The significance of these structures is unknown, although the cap is probably a nucleolus.

Besides the food vacuole there are one or more conspicuous contractile vacuoles which continually empty out to the exterior. These vacuoles are present, as well, in the aggregative and migration stages and can be seen easily in living preparations (Plate 1).

The size range of the cells is greatest during the vegetative stage. If the cells are rounded up by placing them in a drop of standard salt solution, it is possible to measure the diameters (Bonner and Frascella, 1953); they range from about 5.5 μ to 16.0 μ, which means a difference of over twenty fold in volume between the smallest and largest vegetative cells. This matter was also checked using the more reliable method of measuring the diameters of the nuclei stained by the Feulgen method (Fig. 9), where a range from 2.6 to 5.2 μ was found, an eight-fold difference in volume between the smallest and largest nuclei (Bonner, Chiquoine, and Kolderie, 1955).

The aggregation stage of *D. discoideum*

There is a considerable period, lasting from four to eight hours, between the vegetative stage and the aggregation stage. The appreciation of this interphase has come recently, especially from cytological studies in which it became obvious that the cells cease feeding and change their staining charac-

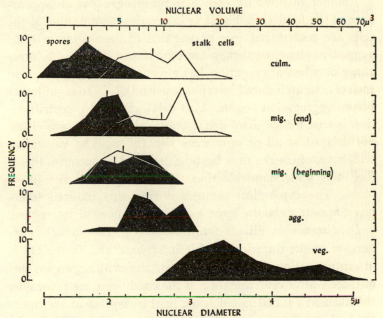

Fig. 9. The distribution of nuclear size measured at different stages of development in *Dictyostelium discoideum*. Each sample is based on the measurement of thirty nuclei. *Culm,* culmination; *mig,* migration; *agg,* aggregation. (From Bonner, Chiquoine, and Kolderie, 1955.)

teristics. A difference in staining properties between vegetative and aggregating amoebae was first noticed by Arndt (1937), who reported a dark refractile body, as revealed by iron haemotoxylin, that appears only during the later stages. It is difficult to know what this corresponds to in our studies, but there are certainly differences in staining between the two stages. The most striking is the reappearance of the metachromatic granules during this interphase period. First a few and then progressively more cells will show this property, and we were able to demonstrate that metachromasia appears before the production of or the sensitivity to the chemotactic substance acrasin.

Another obvious difference is the progressive disappearance of the food vacuoles. It is interesting to note that if the cells are centrifuged free of bacteria at any point in their vegetative phase, or if they are centrifuged during the interphase or when aggregation has already begun, there will in each case be an induced interphase period of six to eight hours before aggregation begins. This delay caused by centrifugation is true for *D. discoideum* only; other species are either not delayed at all or only very slightly so. The vegetative cells have, of course, now become deprived of bacteria, therefore it is understandable that they should enter the interphase. The cytological changes during this induced interphase appear to be the same as those of a normal interphase.

D. discoideum differs from *D. mucoroides* or *D. purpureum* in the duration of this interphase and in the effect of centrifugation. The latter two species will aggregate immediately after centrifugation if the conditions are favorable.

Takeuchi (1960) noted some further aspects of the interphase that are important. There is a particular point in the interphase where a number of changes occur more or less simultaneously: the rate of movement doubles (a matter pursued in greater detail by Samuel, 1961, and which will be discussed in the chapter on morphogenetic movements); the mitochondria change from dot-like structures to filaments, the succinic dehydrogenase activity in the cells increases while the cytocrome oxidase shows a slight decrease.

Another striking change that occurs in the interphase is a radical decrease in cell size. The mean diameter of an actively feeding vegetative cell is about 11 μ, while the mean of an aggregating cell is 7 μ (Bonner and Frascella, 1953). This great difference is no doubt accounted for largely by the fact that during the interphase period, when feeding ceases, there is a gradual disappearance of the many food vacuoles. By the more accurate method of measuring the size of the nuclei, it is found that the vegetative nuclei have an average diameter of 3.6 μ, and the aggregating nuclei, 2.5 μ, indicating that the nucleus during aggregation averages two thirds the volume of the nucleus of the vegetative amoeba (Fig. 9).

At the end of the interphase period the cells become more elongate; this is a striking change, noted by many of the early workers (Plates 3, 4). The pattern of the first signs of aggregation will depend to a great extent both on the concentration of the amoebae over the substratum and on the amount of water about the amoebae. If the amoebae are under water they will frequently first form large sweeping streams that extend 2 or 3 cm or more without any real center (preplasmodium). This pattern soon breaks up into the more usual one of small centers that start with a mere four or five amoebae and eventually spread to an area ½ cm in diameter. Initially, the amoebae come in separately, directly to the center, but very soon they tend to flow together into streams. From Arndt's (1937) time-lapse motion pictures of *D. mucoroides* and Bonner's (1944) of *D. discoideum,* it is evident that the amoebae rarely flow in smoothly but usually aggregate in pulses; that is, there is a wave of fast inward motion that spreads outward like the ripple produced by a stone tossed into a still pond.

D. discoideum is unique in that the centers do not all appear simultaneously and often occur in the middle of large streams of an early, large aggregate (Gerisch, 1961a; Bonner and Hoffman, 1963).

Shaffer (1956 et seq.) has examined the aggregation process in great detail and has made many new and valuable observations. Most of them bear on the mechanism (including an interpretation of the possible significance of the pulsations) and will therefore be dealt with in more detail in Chapter IV, although one observation is pertinent here. Earlier workers noted that the cells appear at some moments to fuse in a stream and at other moments to separate. Shaffer finds that the amoebae can exist in two main phases: one in which they adhere to one another little or not at all, and another in which they adhere strongly. Those in the second phase (which represents a more advanced stage of development) may be said to be "integrated," to use Shaffer's term. Frequently an integrated stream of cells will spontaneously break up or "disintegrate," only to reaggregate and reintegrate elsewhere. Arndt (1937) makes the point that centers may often begin to form and then dissolve, and Shaffer not only concurs but finds that the normal process of development frequently involves a series of successive stages of integration and disintegration, integration finally taking ascendency in the formation of a pseudoplasmodium. The haphazard nature of this backward and forward progression emphasizes, according to Shaffer, the instability of this primitive associative development; it is as though integration-disintegration teeters delicately on a fulcrum, the advantage on the integration side sometimes being very slight indeed. In some recent work (Bonner and Hoffman, 1963) there is evidence that the disintegration may be due to an inhibiting substance produced by neighboring centers and can be prevented by adding adsorbents, such as charcoal, to the culture dishes.

Occasionally the aggregation streams do not come toward a solid center, but form a whirlpool, giving rise to a center shaped like a doughnut (Arndt, 1937; Raper, 1941b). These unusual centers are totally absent, however, if charcoal is present in the dish. Thus it would seem that center dissolution and these whirlpools are both affected by some substance given off by the cells.

During the interphase period before aggregation there are at first very few and finally no cell divisions (Bonner and Frascella, 1952). Unfortunately, this information is based only on the interphase induced by centrifugation, for the normal interphase is obscured by the presence of bacteria. This absence of mitoses continues until the latter part of aggregation, and then some divisions appear, first in the apical papilla that rises above the agar, and finally in the whole mass during migration. However, there is a demonstrable difference in the distribution of mitoses in the front (prestalk) and hind (pre-spore) regions, which we suggested might reflect an early sign of differentiation. These observations, along with those of Wilson (1952 et seq.), emphasize the fact that the previous supposition that all cell division ceases at the aggregation phase is incorrect, and that aggregation only separates the feeding from the non-feeding, morphogenetic stages. These divisions are not especially numerous, the maximum being about one division per 100 cells for a short period immediately following aggregation, falling to one per 200 after ten hours of migration (Bonner and Frascella, 1952).

There are a number of points of sharp disagreement between our work and that of Wilson. He concludes that there are seven chromosomes, and we suggested four, with seven arms. Labudde (1956) concurs with Wilson, and seven is no doubt correct.[3] Whether arms or whole chromosomes, the number is uneven, which means that this is the haploid complement.

Another difference between our findings and Wilson's is that we interpreted the chromosome figures to be mitotic, and he interprets them as meiotic. This brings up the whole subject of his and Skupienski's (1920) views on sexuality in the Acrasiales, which will be briefly discussed.

As previously mentioned, both authors have suggested,

---

[3] Skupienski (1920) also states four chromosomes, but his illustrations are such that his conclusion has little weight. He did not have the advantage of orcein smears. E. W. Olive (1902) shows three chromosomes in his figures, but again no significance can be attached to this observation. Both of these authors worked with *D. mucoroides*, but seven is known to be the number there also.

contrary to the negative evidence of all other workers, that amoebae fuse in pairs at the onset of aggregation, and that meiosis appears subsequently in the pseudoplasmodium.[4] Aggregation is, according to Wilson, the diplophase. In Skupienski's scheme the aggregate is made up entirely of zygotes which then undergo two reduction divisions so that each zygote nucleus gives rise to four haploid spores. Wilson also states that "aggregation may be looked upon as a congregation of individuals, and each individual carries on the process of syngamy and meiosis according to the time of its arrival." In Wilson's scheme there are two meiotic divisions during and immediately following aggregation, and then two divisions just prior to stalk formation, so that each zygote produces sixteen spores.

It is possible now to state that both these proposed cycles are probably incorrect, and sexuality remains, as before, undemonstrated. There are many points in which the proposed cycles are inconsistent with the observations of other workers, as has been discussed by Sussman (1955b, 1956b), Raper (1956a), Shaffer (1958, 1962), Huffman and L. S. Olive (1963).

In the first place the process of aggregation has been examined in living preparations by numerous workers, and a careful time-lapse motion picture study was made by Huffman and Olive (1964). They present evidence that no true syngamy occurs, but only engulfment or cannibalism.

Another difficulty arises when Wilson proposes that cells of large size are zygotes. From studies on the sizes of cells

[4] Skupienski (1920) further argues for a plasmodial stage, but in this he stands entirely alone. The evidence from sectioned material, as well as from many other observations, has definitely ruled this out. The only observation remotely similar to this is the recent discovery by Huffman, Kahn, and L. S. Olive (1962: see also Huffman and Olive, 1964) that the amoebae will show temporary fusions or anastomosis, especially in the period just before aggregation. Also, of course, there is the fact that frequently bi- and even tri-nucleate cells are found.

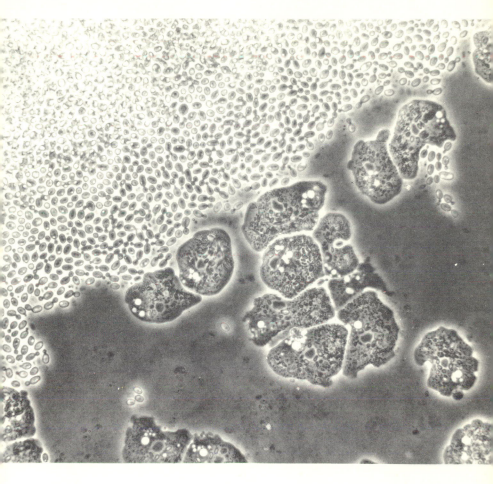

PLATE I. A group of amoebae of *Acrasis rosea* feeding on a yeast (*Rhodotorula* sp.) (Photograph by K. B. Raper.)

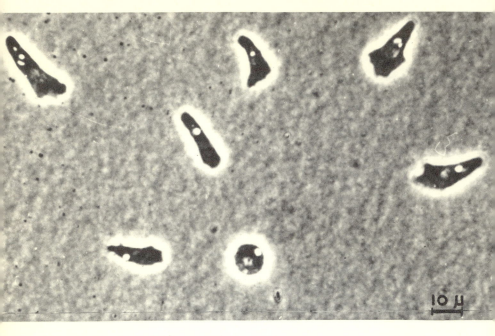

PLATE 2. Founder cells. *Above*: a founder cell beginning aggregation in *Polysphondylium violaceum* (photograph by B. M. Shaffer). *Below*: A founder cell in *Dictyostelium minutum* (photograph of G. Gerisch).

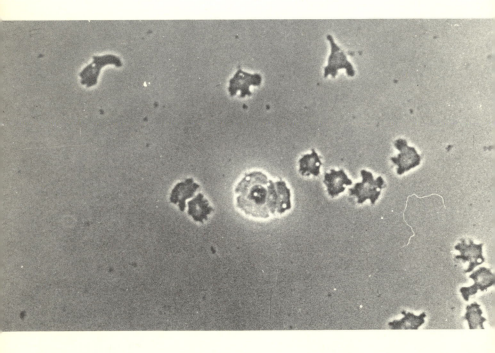

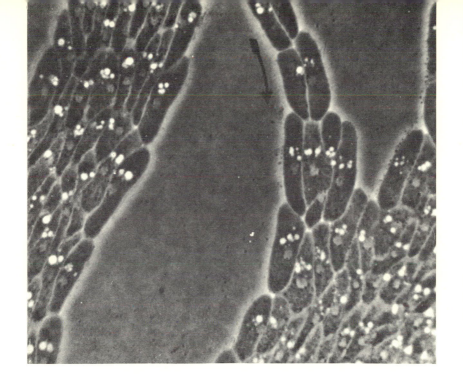

PLATE 3. Aggregation in *Dictyostelium discoideum*. *Above*: highly magnified section of a stream. *Below*: an entire aggregate. (Photographs by K. B. Raper.)

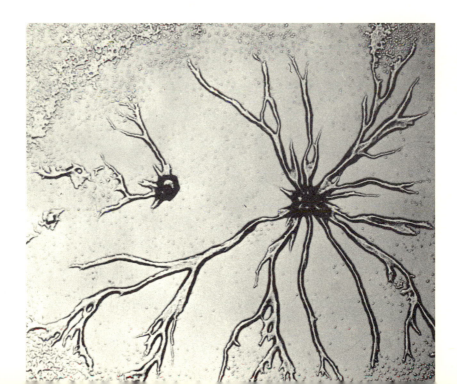

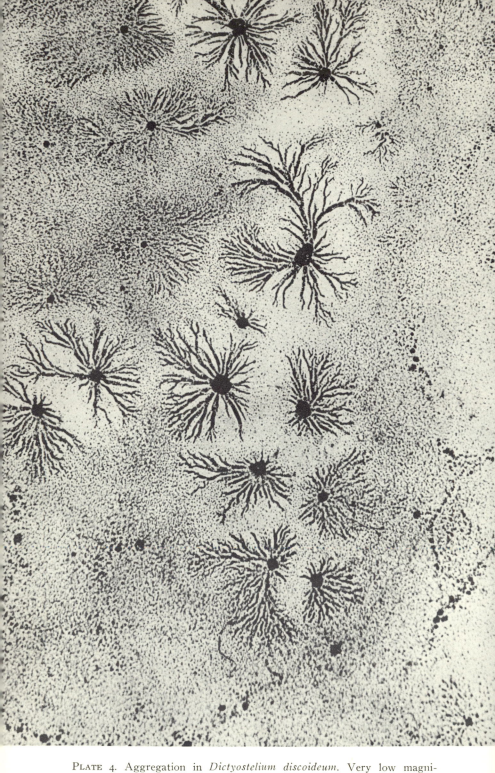

PLATE 4. Aggregation in *Dictyostelium discoideum*. Very low magnification to show many aggregates. (Photograph of K. B. Raper.)

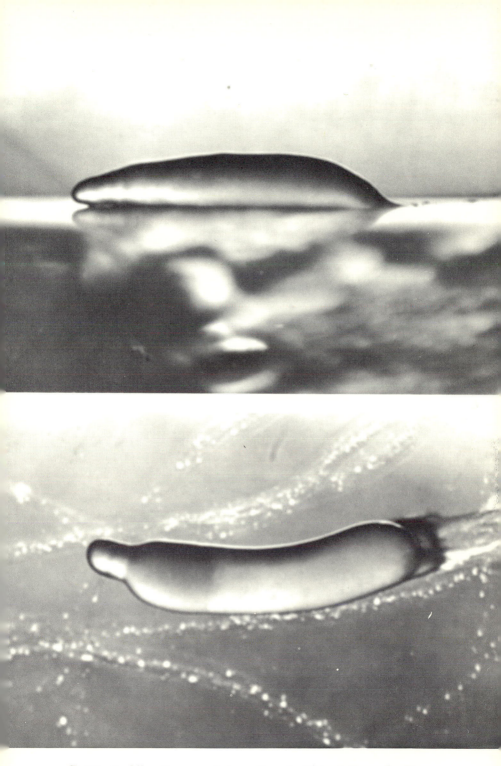

PLATE 5. Migrating pseudoplasmodia of *Dictyostelium discoideum.*
Above: A side view. *Below*: A view from above. (Photographs by D. R.
Francis.)

PLATE 6. Culmination of *Dictyostelium discoideum*. Each photograph represents a time interval of approximately one and one half hours.

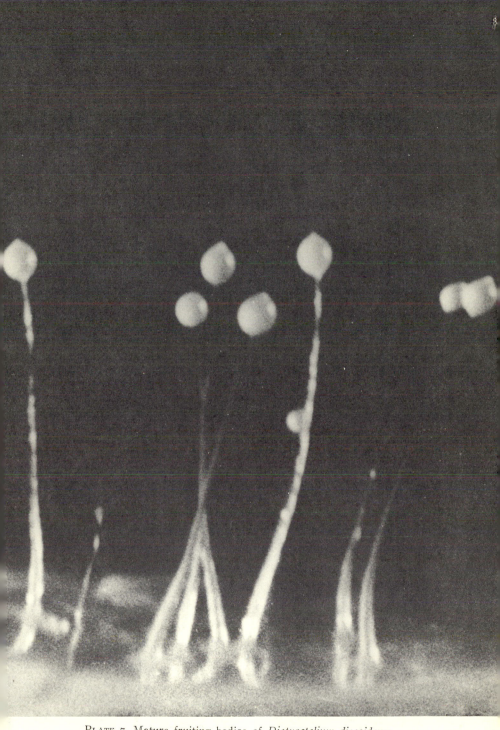

PLATE 7. Mature fruiting bodies of *Dictyostelium discoideum*.

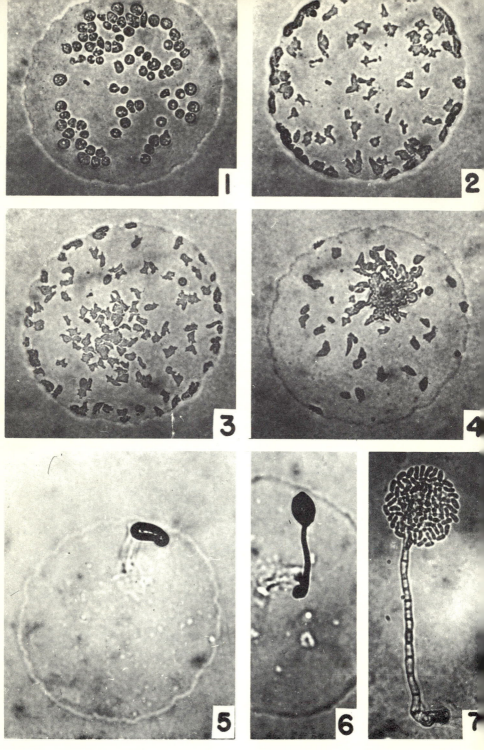

PLATE 8. Composite photograph of the complete morphogenesis of *Dictyostelium discoideum* taking place in a small confined drop on an agar surface. 4, aggregation; 5, migration; 6, culmination; 7, the mature fruiting body pressed against the agar surface showing the stalk cells and the spores. (Photographs by Konijn and Raper.)

and nuclei (Bonner and Frascella, 1953; Bonner, Chi-quoine, and Kolderie, 1955) it is obvious that this proposal is likely to be misleading, for not only is the size-frequency distribution continuous and not bimodal, but also it extends over a considerable range.

Perhaps the most cogent argument against the above two sexual schemes is that, if one takes into account the time at which they propose meiosis occurs and the number of associated cell divisions during the cycle, then it can be shown that certain fruiting bodies of small cell number could not form as in fact they do. Sussman (1955a) produced a small mutant of *D. discoideum* in which he has obtained fruiting bodies of twelve cells, nine of which were spores. According to Skupienski's scheme, such a sorocarp would have consisted of three cells at the aggregation stage (not confirmed by Sussman), and according to Wilson's scheme there would have to be ¾ of a cell at aggregation. The argument against such meiotic divisions is even stronger for the individual of seven cells found in *P. pallidum* (Bonner and Dodd, 1962a).

The ideal way to settle the issue would be to demonstrate recombination, but unfortunately work along this line is also in an unsatisfactory state. It was first attempted by Sussman (1956b) without success, but later Sussman and Sussman (1961) reported one instance of recombination, which they have not been able to repeat. Rafaeli (1962) has also attempted to find recombination in *Polysphondylium violaceum* but without success. Better genetic evidence is definitely needed.

One interesting observation, first made by Wilson and Ross (1957), is that some cells within the cultures are diploid and contain 14 chromosomes. This matter has been examined in greater detail by M. Sussman and R. R. Sussman (1962) and R. R. Sussman and M. Sussman (1963), and they have shown that some strains are permanently diploid, some haploid, and some contain both types of cell, the proportion depending to some extent upon the culture conditions. There are a number of possible interpretations of this curious phenomenon, and they do not bear critically, as yet, on the question of whether or not true sexuality exists in these forms.

The migration stage of *D. discoideum*

One of the reasons that *D. discoideum* attracted so much attention was the fact that it was the first cellular slime mold shown to have a stalkless migration period. The possibility of making grafts and performing various kinds of cutting experiments was immediately recognized and exploited by Raper with great profit. It has turned out, since, that many of these experiments can be performed with stalked forms, but *D. discoideum* showed the way. Raper (1956b) has since discovered another species with a stalkless migration period, but *D. polycephalum* has not yet made any contribution as experimental material.

Again, as with aggregation, the details of the movement must wait for a later chapter, but here let me say briefly that where the aggregating streams merge, the rising, cone-shaped center turns directly into a migrating pseudoplasmodium. This now leaves the site of its formation with a slow, gliding movement. Sometimes the tip lies flat along the agar, while at other moments it may be raised, and it was noticed (Bonner, 1944) that when touching the substratum the tip moves more slowly, but when free in the air it shoots forward; it resembles the human tongue in its ability to assume different shapes (Figs. 8, 16; Plate 5).

During this stage the cell mass deposits a thin slime sheath which, like a collapsed sausage casing, lies flat on the surface of the substratum behind the advancing pseudoplasmodium. There is no evidence from the histological work that any particular group of cells, either near the surface or in the interior, is responsible for secreting this slime sheath and therefore it has been presumed to be excuded by all the amoebae. It can be easily shown, by plunging a migrating mass under the surface of some water, that the slime sheath is probably thinnest at the anterior end, for the cells will disengage themselves from the mass first in that region in the unfavorable aqueous environment. It can also be shown that the slime track itself does not move, but that the amoebae move within it. This may be readily done by placing a marker on the surface of the sheath (either carbon black or *Lycopodium* spores); the marker will retain a fixed position, while the pseudoplasmodium will slip out from under as it moves forward.

There have been some attempts to identify the slime sheath material chemically, but this has proved difficult because of its extreme thinness. Results with staining techniques for cellulose led Raper and Fennell (1952) to suggest that the slime sheath is made up partly of cellulose and partly of mucin. The evidence for cellulose comes from a positive reaction with chloroiodide of zinc, although Schweitzer's reagent does not destroy the slime track as it would were it pure cellulose. But these authors found that the birefringence, which is normally demonstrable under polarized light, is obliterated by the Schweitzer's reagent. However, the more recent electron microscope study of Mühlethaler (1956) shows that there are no fibrils in the sheath, which he interprets as positive evidence for the complete absence of cellulose.

In general, in the stained sections of the migrating stage the cells are somewhat rounded, lacking the elongate orientation characteristic of the aggregation stage. There is evidence of pseudopods, often interlocking between cells, but they are ordinarily not extended to any marked degree. The exception to this usual picture is at the very beginning of migration. Frequently at that stage, the cells can be seen to form a whirlpool or some kind of curved orientation within the mass. Also in some cases where there is a narrow tapering apex of an early migrating pseudoplasmodium, the cells will show elongation in a transverse direction, perpendicular to what must have been the axis of thrust. This orientation is seen again in the culmination stage (Figs. 8, 16).

One of the principal objects of the various histological studies was to detect early signs of cell differentiation, for it was known from the grafting experiments of Raper (1940b) that the anterior end of the migrating mass gives rise to the stalk cells, and the posterior end gives rise to the spore cells. He showed this by growing *D. discoideum* on the red bacterium *Serratia marcescens*; since the amoebae retained the pigment prodigiosin, it was possible to obtain red pseudoplasmodia. Then by grafting a red anterior onto a white posterior of a migrating mass (and vice versa), he was able to follow the fates of the different regions; in the words of the experimental embryologist, he mapped out their prospective significance.

In some early studies using haematoxylin, it was easy to see that at the beginning stages of migration there was no difference in staining properties between the anterior and posterior cells, but that at a later stage the two regions were clearly identifiable and were separated by a sharp division line (Bonner, 1944). The differences were of two sorts; the posterior cells stained more darkly with the haematoxylin, and they were smaller than the anterior cells.

The size difference has been pursued both by directly measuring the cell diameter in living preparations (Bonner and Frascella, 1953) and by the more satisfactory method of measuring the diameters of nuclei which have been prepared by the Feulgen method (Bonner, Chiquoine, and Kolderie,

1955). In a late migration stage the mean diameter of the posterior cells (rounded by being separated and placed in a standard salt solution) is 7.7 μ, while that of the anterior cells is 8.5 μ; the difference between the two is highly significant statistically despite the great overlap of their ranges. In the study on nuclear size, there was no demonstrable difference between the front and hind cells at the beginning of migration; both had a mean diameter of approximately 2.1 μ. But in a late migrating pseudoplasmodium the posterior cells had a mean of 1.9 μ and the anterior cells a mean of 2.6 μ; again, despite the large range, these differences are highly significant. One further point here is of considerable interest. As can be seen from Fig. 8, at the beginning of migration the cells in the posterior region (the presumptive spore cells) have smaller nuclei than those of the aggregating cells, and the nuclei of late-migration presumptive spores are still smaller. In fact, a glance at the whole figure shows a continuous decrease in size from the vegetative stage to the encapsulated spores. The presumptive stalk cells, on the other hand, show an increase in nuclear size at late migration; therefore, the difference that appears between the two regions during the migration stage is due in part to a decrease in the size of the nuclei of the posterior presumptive spores, but to an even greater extent to the swelling of the anterior presumptive stalk nuclei.

The difference in the staining reaction of the two regions with haematoxylin could not, however, be accounted for by the change in cell size. The presumptive spore cells contained small, round, deeply stained granules that were totally absent in the presumptive stalk cells. Still another difference in the staining properties was discovered quite by accident. It was found that a number of vital dyes, Bismarck brown, neutral red, Nile blue sulphate, and even the prodigiosin of *Serratia marcescens*, would produce, at first, a uniformly colored migrating pseudoplasmodium, and then at a later stage the anterior end would remain dark while the posterior end blanched, thus giving a vital demonstration of the two presumptive regions (Bonner, 1952).[5] The dye was incorporated into the amoebae at the vegetative stage, the excess washed free by centrifugation, and the stained amoebae placed on non-nutrient agar so that aggregation and migration could follow. The blanching of the posterior end was recorded by time-lapse photography, and it was possible to show that the change took place rapidly, in a matter of ten to fifteen minutes. This rapid change excluded the idea that there had been a movement of the dye from one part of the pseudoplasmodium to another, and indicated, rather, that there must have been a change in the condition of the dye in the cells to give this blanching. Another argument in favor of an alteration of the dye within the cells comes from cutting experiments, in which an all-blanched segment will produce a new dark tip, and an all-dark segment will produce a blanched posterior end. These coloration changes are therefore strictly reversible. The only unfortunate part of this story is that we have no idea, as yet, of the nature of this dye change. The fact that so many different dyes show the effect would suggest that there is some sort of non-specific clumping phenomenon of the dye particles, or perhaps an oxidation-reduction effect, but there is no evidence on this point.

With the idea of obtaining some further insight into the differences between the presumptive stalk and presumptive spore areas, a program was launched using some of the more modern histochemical techniques (Bonner, Chiquoine, and Kolderie, 1955). A number of different methods were tried, but the results of one in particular will be described here—

[5] By far the most effective vital dye is cresyl violet, a fact which was discovered by Takeuchi (unpublished).

the periodic acid–Schiff reaction for polysaccharides.[6] The vegetative and the aggregating amoebae show numerous granules containing polysaccharide, and at the end of aggregation the cell mass stains homogeneously. As migration proceeds, the division line between pre-stalk and pre-spore is sharply delineated by this PAS reaction (Fig. 10). The posterior pre-spore cells are now laden with numerous small, darkly pigmented, buckshot-like granules. These correspond closely to the granules revealed by haematoxylin, indicating that the differences between the two regions demonstrated by the earlier haematoxylin studies were differences in the non-starch polysaccharides. These differences become even more exaggerated as culmination follows.

During the migration phase the PAS preparations show a small wedge of cells at the very posterior end of the mass which show the staining characteristics of the anterior pre-stalk cells. Furthermore, it is obvious that some of these rear-guard cells keep falling behind and may be observed in the slime track, where they cease all forward movement and appear somewhat vacuolate. As will be shown in the discussion of culmination, there is considerable evidence indicating that these rear-guard cells give rise to the basal disc.

The culmination stage of *D. discoideum*

At the end of migration the tip of the cell mass ceases movement while the posterior portion continues to gather in, thereby causing the pseudoplasmodium to round up. The hind end shoves itself under the main mass, and the principal axis now becomes vertical (Fig. 8). The whole mass then flattens itself on the surface of the substratum, as the motion pictures and time-interval photographs show, only to rise again soon and continue its upward movement, the actual culmination. As this happens, the presumptive spore mass is pulled away

---

[6] The preparations were digested in saliva, therefore we were staining primarily for the non-starch polysaccharides.

Dictyostelium discoideum

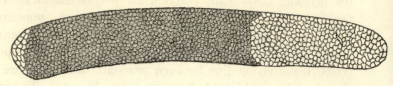

D. mucoroides or *purpureum*

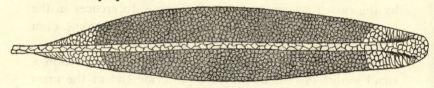

Polysphondylium

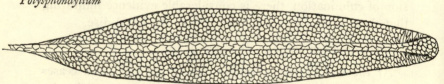

Fig. 10. The staining properties of the cells (with the periodic acid–Schiff method) during the migration phase for three different types of cellular slime mold. (From Bonner, 1958.)

from the surface of the substratum and lifted into the air; the apical papilla of presumptive stalk cells also rises, sometimes smoothly, but more often in throbbing pulsations. This papilla of pre-stalk cells slowly diminishes in size until finally the spherical or more often lemon-shaped sorus lies at the tip of the delicately tapering stalk (Plates 6, 7).

If these changes are followed internally, primarily with the use of stained sections (Bonner, 1944), it is possible to see the exact fate of the two cell types. The presumptive stalk cells first turn into true stalk cells in the apical region by a rounded group of cells becoming vacuolated. Further stalk cells are continuously added to the group from above, in this way pushing the first-formed mature stalk cells downward like a wedge toward the substratum, right through the cell mass. The fact that this occurs, rather than the stalk cells arising *in situ* up and down the main axis, which had been the suggestion of earlier workers, was conclusively proved using Raper's (1940b) technique of grafting a red pseudo-plasmodium tip onto a white posterior portion (Bonner, 1952). In this particular preparation one can see clearly that the initial stalk cells are entirely red, and are pushed down through the white pre-spore cells (Fig. 11). The flattening of the whole mass, mentioned above, occurs when the new stalk is pushing its way down; when it reaches the substratum it meets with some resistance for the first time, and now the piling of the presumptive stalk cells on top of the rigid stalk (by a reverse-fountain movement) results in the rising of the pre-stalk and pre-spore mass.

Around the base of the stalk there are the rear-guard cells, which we noted have pre-stalk polysaccharide staining characteristics. These cells now become vacuolated and directly produce the basal disc. As the pre-spore mass rises, there may be some remaining rear-guard cells attached at its posterior end, indicating either that all of them did not become basal disc cells, or that they accumulated in that region after the formation of the basal disc.

From the histological sections it is clear that final spore differentiation begins on the upper edge of the pre-spore mass and progresses inward and downward from that region. There is a mild controversy as to whether this occurs quickly

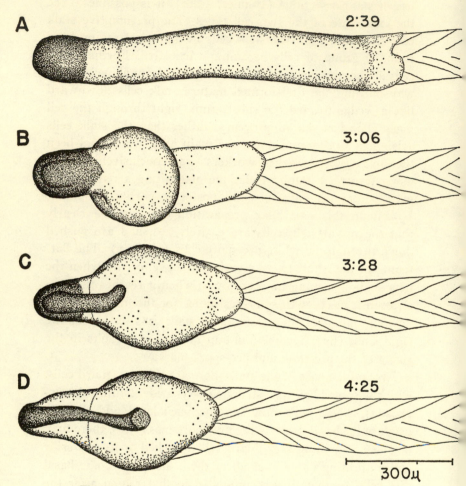

Fig. 11. Camera lucida drawings (surface view) showing how the stalk is first formed at the tip and is pushed downward through the pre-spore cells to the substratum. The dark tip was obtained by grafting the tip of a colored migrating cell mass onto a decapitated colorless one. (From Bonner, 1952.)

(Bonner, 1944) or slowly (Raper and Fennell, 1952). My evidence is based entirely on the correlation between the appearance of opacity of the sorus in motion pictures and serial photographs and the final encapsulation of the spores as revealed in stained paraffin sections; if the correlation is sound, then the process occurs in less than an hour. The discrepancy between our views might be caused by the presence of the rear-guard cells which do not differentiate into spores.

To return to the matter of stalk formation, there are a number of further points worth considering. The stalk sheath itself is a cylinder of cellulose that lies outside the cell walls of the stalk cells proper (Raper and Fennell, 1952).[7] There is a delicate taper to the stalk, and the problem of what factors govern this taper has been raised by Harper (1926), who imagined that the decreasing weight of the sorus caused by a pressure stimulus would affect the diameter of the stalk. There are many objections to such a notion, one being that a tapered stalk will form in inverted cultures where the fruiting bodies are pointing downward. The more reasonable view is that expressed by Raper and Fennell (1952), that the mass of pre-stalk cells diminishes as culmination proceeds, and that the progressive taper comes with the reduction of the diameter of the pre-stalk mass. This matter of taper, as the observations of Pfützner-Eckert (1950) on the taper of *D. mucoroides* indicate, remains an unsolved problem of considerable interest.

The question of what group of cells contributes the stalk material was answered by our studies with the polysaccharide stain. The pre-stalk cells, as was well known, become elongate and line up with the long axis approximately perpendicular to the stalk axis. The inner portion of these cells is extremely rich in non-starch polysaccharide, giving a clear indication

---

[7] This can be seen particularly well in an electron micrograph of Gezelius and Rånby (1957, Fig. 10).

that they are actively secreting the stalk sheath. The actions of any particular pre-stalk cell then have the following sequence. As the cell rises into the upper portion of the papilla, it elongates and becomes a part of a transitory columnar epithelium. After secreting its polysaccharide as a contribution to the stalk sheath, it passes up to the tip and becomes incorporated into the stalk proper. By a gradual process of enlargement it becomes vacuolated and finally secretes its own cellulose cell wall.

The fact that the stalk sheath consists of cellulose was first suggested by Brefeld (1869). This was supported by the work of Olive (1902) and Raper (1940a), but the first comprehensive attack on the problem is that of Raper and Fennell (1952). By the use of chemicals as well as X-ray diffraction studies, paper chromatography of hydrolysis products, and decomposition by cellulose-destroying bacteria, they established the cellulose nature of the stalk sheath beyond question. The X-ray diffraction patterns were those of a hydrate cellulose, and this and other points were corroborated by studies of Mühlethaler (1956). Furthermore, Mühlethaler showed with electron micrographs that the fibrils in the sheath are beautifully oriented in a parallel array, while the fibrils in the walls of the stalk cells are oriented at random. The diameter of the fibrils is the same for the stalk sheath, the stalk cells, and the spore casing: about 70-100 Å, which is, according to Mühlethaler, thinner than the fibrils of the cells of higher plants. In a separate study Gezelius and Rånby (1957) agree, for the most part, with these observations, although they consider the cellulose to be partly mercerized.

When the cells first enter the stalk proper they are small, but as newly arrived cells pile on top of them they expand rapidly. It is impossible, because of the irregular shape of the cells, to make an accurate estimate of cell volume and so gauge the extent of this increase in volume. Furthermore,

in this instance our studies on the nuclear size were of no
help, because apparently the increase in cell size is not ac-
companied by a corresponding change in the nucleus. There
is, in fact, no appreciable change in the nuclear size between
the pre-stalk cells at the end of migration and the young stalk
cells (Fig. 9). When vacuolization of the stalk cells becomes
pronounced at about the level of the center of the spore mass,
then the nuclei become irregular and crinkled and no longer
appear healthy. This suggests that mature stalk cells are for
all intents and purposes dead cells, a fact which is fully confirmed in the detailed studies of Whittingham and Raper
(1960).

13. *Dictyostelium polycephalum*

This interesting species, discovered by Raper (1956b),
has a number of features which mark it as quite different
from other members of the genus (Fig. 12). Raper made a
careful study of its life cycle (1956b), and I shall indicate
some of the principal points here.

To begin with, the germination process is unusual. Instead
of a longitudinal split down the side of the elliptical spore
case, the spore wall bulges out around the equator, where it
eventually bursts, liberating the protoplast and leaving either
two separate or two partially attached, empty, hemispherical
spore cases.

The vegetative stage of *D. polycephalum* is similar to that
of other members of the Acrasiales, while the aggregation is
different in that it lacks the conspicuous thin radiating
streams; rather, there are broad sheets of cells that converge
on the central collection points. Each of these central masses
will give rise to as many as a dozen or more migrating
pseudoplasmodia.

The migrating masses are stalkless as in *D. discoideum,*
but here the similarity stops. They are extremely thin and

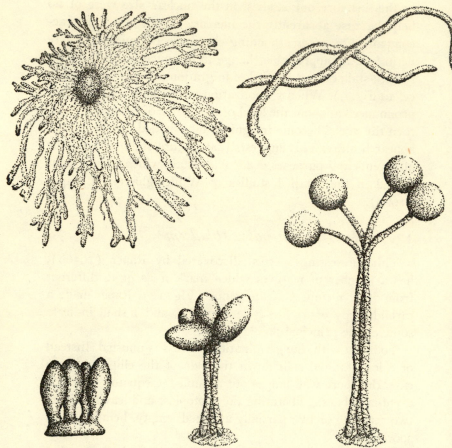

Fig. 12. *Dictyostelium polycephalum. Upper left,* an aggregation pattern; *upper right,* migrating pseudoplasmodia; *bottom,* progressive steps during culmination beginning with one pseudoplasmodium breaking up into a series of small fruiting bodies. (Drawings based primarily on photographs by Raper, 1956b.)

long—up to 5 to 10 mm. in length and about 50 to 60 μ in diameter. These string-like masses have the further peculiarity of showing no tropisms to light and heat.

From some material kindly sent to us by Raper we made some periodic acid–Schiff preparations for non-starch polysaccharides of these migrating masses and found them to show no signs of early differentiation whatsoever. This is not surprising considering the nature of the subsequent development.

The migration process ceases when the pseudoplasmodium accumulates in one large globular mass; this mass then becomes subdivided into a series of one to nine papillae, each of which initiates stalk formation and rises as a fruiting body. As these culminate, their stalks fuse or lie alongside one another, cemented together for at least three-quarters of their length, after which each stalk juts out, with a spherical sorus at its apex. The result is a coremiform fructification, in which there are a series of *D. discoideum*-like fruiting bodies (minus the basal discs), bound together along most of their stalks like a bunch of flowers in a narrow vase (Fig. 12). It would be most interesting to know if the point at which each individual fruiting body diverges from the mass is also the point at which final spore differentiation occurs. This would seem to me a reasonable possibility, because from this moment on, as a result of spore encapsulation, there would be a negligible production of the binding slime sheath material. This would also argue that the main cohesive force in bringing the stalks together is in the adhering sheath of each sorocarp.

The conditions which favor fructification in this particular species are somewhat different from those encountered in other forms (Whittingham and Raper, 1957). In the first place, the optimum fruiting temperature is about 30° C (with a maximum at 34°-35° C), while other species of *Dictyostelium* have an optimum closer to 22°-24° C and a maximum at approximately 30° C. Also it was found that *D. polycephalum* fruited with far greater frequency in a three-membered culture with a mold (*Dematium*) as well as the bacterial food supply (Raper, 1956b), and in the more recent study Whittingham and Raper were able to imitate the advantageous effect of the *Dematium* by slightly lowering the relative humidity of the atmosphere.

14. *Polysphondylium violaceum* and *P. pallidum*

The genus *Polysphondylium* is easily recognized by the whorls of branches that jut out from the mature sorocarp (Fig. 6). The common *P. violaceum* is a large and conspicuous form, although again there is considerable variation among isolates, especially as to the duration of migration (Bonner and Shaw, 1957). The less common, white-spored *P. pallidum* is much smaller and more delicate in its appearance. As is true of small species of Acrasiales in general (especially of *D. lacteum, D. minutum,* and *Acytostelium*), *P. pallidum* grows poorly on rich nutrient media and does best where the bacterial population is kept relatively low (Raper, 1951).

From spore germination to the early migration stage *Polysphondylium* is very similar in appearance to *D. mucoroides.* Shaffer (1957a) has pointed out some differences in the aggregation patterns, but the morphological changes are basically similar.

As culmination proceeds, a group of cells is pinched off from the posterior end of the pseudoplasmodium, and this detached mass, which surrounds the stalk like a doughnut, subdivides into a series of small fruiting bodies that jut out from the stalk at right angles. The process is repeated many times, until finally the whole fruiting body has the appearance of a miniature Christmas tree with a sorus at the tip of each branch and one large terminal sorus at the end of the main stalk (Fig. 6).

Harper (1929, 1932) has made a detailed study of *Polysphondylium* and taken great pains to measure the number and distribution of whorls of a large number of fruiting bodies. On the basis of these measurements he argues for a fixity of structure, despite the great variation in size; that is, the number of whorls and the average number of branches per whorl are directly correlated with the size of the fruiting body. He also found, as did Potts (1902) observing *D. mucoroides,* that the fruiting bodies of *Polysphondylium,* when grown in the light, were smaller than those raised in the dark. The difference, according to Harper (1932), is that the fruiting bodies in the dark had twenty-one per cent more whorls per plant and twenty-three per cent more branches.[8] The effect of light in producing smaller plants is probably a general one for the Acrasiales, as Raper (1940b) points out, for it is found in *D. discoideum* and *D. mucoroides.* It is undoubtedly related to Raper's discovery that light induces aggregation to take place two to four hours sooner. Light therefore induces the early formation of small aggregates.

If a histological analysis is made of the migrating cell mass of *Polysphondylium,* it is possible to see that it differs significantly from *D. mucoroides* (Bonner, Chiquoine, and Kolderie, 1955; Bonner, 1957). *Polysphondylium* is one of those species that produces a stalk which persists through the entire migration stage, yet there is absolutely no evidence of any division line between the pre-stalk and the pre-spore cells. All the cells except those that are either in the stalk or on the verge of entering it have a homogeneous non-starch polysaccharide distribution, showing staining properties similar to those of aggregating cells (Fig. 10).

This embryonic condition is sustained even after the groups of cells are cut off the posterior end prior to their formation of the whorl of branches. Repeated attempts were made to catch the moment when each small branch would show pre-stalk and pre-spore cells, but the stained sections showed

---

[8] These figures are in per cent of branched fruiting bodies only, discounting those that had no branches at all.

either no differentiation or else final differentiation of the spores; the transition period must be very short indeed.

The fact that *Polysphondylium* has such a radically different pattern of differentiation from *D. discoideum* and *D. mucoroides* argues strongly for the usefulness of the comparative method in the experimental analysis of these forms. For instance, we had presumed from the earlier work on *Dictyostelium* that no stalk could be produced without a pre-stalk zone, but *Polysphondylium* does this normally. There is, of course, also a sound rationalization as to why *Polysphondylium* has this delayed differentiation. After an extended period of migration the cell mass cuts off groups of cells at the posterior end and these groups further subdivide into what are essentially a cluster of small sorocarps. In this case the individual fruiting body with a single stalk and a single sorus does not become isolated from the communal cell mass until just before final differentiation. It is therefore not surprising that differentiation should wait until after the units are carved out. As a matter of fact, the same argument would apply to *Dictyostelium polycephalum*, for there, also, there is no differentiation during the migration stage, as was shown by the periodic acid–Schiff preparations, but only after the cell mass has broken up into the units that will be the final discrete fruiting bodies. In both cases there has been a prolongation of the embryonic characters, so that at a late moment the large cell mass can break up into smaller sub-units.

This property, which might be called a special slime mold form of neoteny, emphasizes the striking similarity between *Polysphondylium* and *D. polycephalum*. They differ primarily in the absence of the stalk during migration in *D. polycephalum*. The result is that when *D. polycephalum* breaks up into sub-units, it does so on the surface of the substratum rather than along a stalk as in *Polysphondylium*. Consequently, the fruiting body of *D. polycephalum* is one whorl jutting up from the surface of the substratum. It is true that *Polysphondylium* cuts off whorls at repeated intervals, and there is no equivalent for this in *D. polycephalum*. However, the fact that the doughnut-shaped cell mass does produce a whorl can probably be explained mechanically on the simple basis that it surrounds a cylindrical stalk, and the small fruiting bodies that are the branches tend to come off at right angles. In other words, both forms have the same mechanism of orientation of the fruiting bodies, since in *D. polycephalum* they come off at right angles from a flat substratum.

My argument, then, is that if there should arise a mutant of *Polysphondylium* that lacked the ability to form a stalk during migration, the immediate result might be *D. polycephalum*; similarly a stalkless mutant of *D. mucoroides* might give rise to *D. discoideum*. This hypothetical suggestion brings up an analogous point, namely the relation between *Dictyostelium* and *Polysphondylium* in general. A number of authors have been impressed by the fact that the presence or absence of branches is a very slight difference. This is because small individuals of *Polysphondylium* lack branches, and many strains of *Dictyostelium mucoroides* show marked irregular branching which to some degree indicates an intermediate condition between the two genera (Huffman and L. S. Olive, 1963; Rai and Tewari, 1963a, b). The difficulty in drawing a sharp line between these genera was recognized long ago by E. W. Olive (1902), who said "it is possible that the distinction is not great enough to warrant the retention of the two forms as distinct genera."

15. Summary

There is a great temptation to indulge in phylogenetic

speculation, but unfortunately there is so little basis for it that it soon becomes hollow and unrewarding. Were there a fossil record of the rise of the Acrasiales there might be some point to it, but there is no prospect of any firm facts upon which to hang our hypotheses. We do have, and this is far more important, structural similarities and dissimilarities, and on this basis we can see relationships and understand differences. This says nothing of how they came into being, but only what they are.

On this basis we first made the distinction of the different kinds of amoebae; we divided them into at least four types, the *Hartmanella,* the *Sappinia,* the *Guttulina,* and the *Dictyostelium* types.

Within those species that possess the *Dictyostelium* type of amoebae there are a number of distinctive structural features, some of which involve size:

There are, to begin, the smallest forms that fail to aggregate (*Protostelium*).

There are somewhat larger forms that do aggregate, but, in common with *Protostelium,* the stalks are non-cellular (*Acytostelium*).

A third form, which is similar to the first two in possessing spherical spores, also has a stalk made up of cells, although there may be acellular segments of stalk (*Dictyostelium lacteum*).

All the remaining forms have elliptical spores. They may have a stalk initially, from the end of aggregation, and differ only in size (the *Dictyostelium mucoroides* complex, including *D. minutum*).

They may be highly branched, the branches may be regular, and the spores colorless, that is, white or faint yellow (*Polysphondylium pallidum*).

The stalk may be relatively free of branches and the spores purple (*Dictyostelium purpureum*), or the stalk may be regularly branched with the spores purple (*Polysphondylium violaceum*).

To return to the colorless forms, there may be a period of continuous migration without stalk in a form that does not possess branches (*Dictyostelium discoideum*) or in a branched form where there is a subdivision of the cell mass (*Dictyostelium polycephalum*).

D. discoideum is further characterized by a well-defined basal disc, while the new, undescribed species of Raper (1960a) has a crampon-shaped hold-fast.[9] Finally there is the lost *Coenonia* of van Tieghem which also has a hold-fast base, and an elaborate cup-like sorus.[9]

[9] Recently Raper (personal communication) has discovered a number of other interesting new species of *Dictyostelium* as yet unpublished.

III. Growth

THE remainder of this book will center around problems of development. It is possible to subdivide the elements of developmental processes a number of ways; the system employed here will be one that was devised previously.[1] Development comprises three constructive processes: growth, morphogenetic movement, and differentiation. By growth is meant size increase or the synthesis of new protoplasm; morphogenetic movement is the rearrangement of the cells to make new form; and differentiation is the differences or change in chemical composition and structure both in time and in different parts. Each of these constructive processes is carefully guided and controlled within each organism so that the end result, in terms of size and form, is consistent from generation to generation.

The developmental processes of the cellular slime molds differ in a number of significant ways from those of most multicellular organisms, which begin as fertilized eggs. Some of these differences are of great value in the experimental analysis of development, especially the fact that in the cellular slime molds (more than in most organisms) the growth phase is dissociated from the phases of morphogenetic movement and differentiation, which has the advantage that these processes can be studied separately. In multicellular plants and animals size increase occurs simultaneously with differentiation and morphogenetic movement, and the processes can only be dissociated by experimental means. As Harper (1926) was the first to point out, cellular slime molds grow first and then, with the onset of aggregation, carry on the other two developmental processes.

The line dividing the two stages may not be quite as sharp

[1] Bonner, J. T. *Morphogenesis*, Princeton University Press (Athenaeum), 1952.

as originally imagined, but the basic truth remains. It has been shown that cell division continues into the migration stage (Bonner and Frascella, 1952; Wilson, 1952), although this does not involve any intake of food; clearly feeding has stopped some hours before. However, all digestion does not cease the moment the plate is licked clean; it is some time, for instance, before the food vacuoles containing bacteria disappear (well into the migration stage). It is not surprising, therefore, that some cell division should occur even though there has been no further energy intake for some time.

There is also some doubt as to whether or not differentiation might not begin in a minor way before aggregation. It is known that the cells vary greatly in size prior to aggregation, as was discussed previously, and the question has been raised as to whether or not the cells might differ also in other ways; they might be semi-differentiated before aggregation and be sorted out to some extent during aggregation (Bonner, 1959a; Takeuchi, 1963). This matter will be examined in more detail in the chapter on differentiation. Here we can say that the only firm fact is that there is some sorting out or rearrangement of cells during aggregation, and the cells at the end of the growth phase do differ over a wide range. But whether or not this is the first step in the differentiation process remains an interesting hypothesis in need of more evidence.

The growth phase of the cellular slime molds is unlike the other phases in that it involves single cells and is comparable to that of microorganisms in general, while the later stages of development are multicellular. Because of the aggregation process, the cellular slime molds have the characteristics of two basically different types of organism. It is possible to grow a clone of cells from one spore, and this clone will produce normal fruiting bodies, but it is equally possible to obtain fruiting bodies from the progeny of many spores even though they may possess genetic differences.

We will begin our analysis of the growth phase with germination of the spore, for in a sense this is the beginning— the starting point of the growth process—even though the actual amount of synthesis involved in germination may be very modest. We will then examine the various methods by which the growth of the amoebae has been accomplished in the laboratory, and this examination will be largely an introduction to techniques rather than an analysis of the growth process. The technical difficulties have been so great that we still know exceedingly little about the chemistry of growth in the cellular slime molds. As will be seen, it is only very recently that the first successful cell-free media have been prepared, and they still contain unknown substances, such as serum albumen or embryo extract, and only work with some strains of *P. pallidum* (Hohl and Raper, 1963b,c).

1. Germination

We have already described the mechanics of germination, which generally occurs by the splitting of the cellulose spore case and the escape of a single, uninucleate amoeba. In our analysis of this problem we may consider three aspects of germination: environmental factors which favor germination, the cell-cell interactions during germination, and finally the internal biochemical changes. This three-pronged attack will serve as a model for our discussion of all the developmental processes to be examined.

The temperature and humidity which favor growth also favor germination, as would be expected. The temperature requirements actually cover a wide range, for low temperatures (down to *ca.* 10° C) do not inhibit germination, but merely retard it. The humidity requirements have not been carefully studied, but it is known that humidity must be high, and germination takes place readily under water.

Relatively few experiments have been done on the effects of salts and nutrients upon germination. Potts (1902) suggested that phosphate stimulated germination in *Dictyostelium mucoroides*, but Skupienski (1920) reported good germination in distilled water (as had Cienkowsky, 1873, for *Guttulina*). Russell and Bonner (1960) were unable to show any difference in per cent germination in *D. mucoroides* with and without phosphate buffer.

As far as nutrients are concerned, Russell and Bonner (1960) compared non-nutrient Bacto agar (2 per cent) with

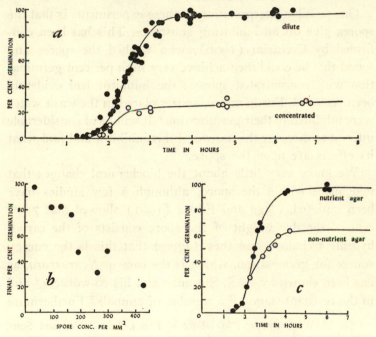

Fig. 13. a) Graph showing the per cent germination at different times for concentrated and dilute suspensions of *D. mucoroides* spores. b) Graph showing the relation of final per cent spore germination to the density of the spores on the agar surface. c) A comparison of the germination curves of spores placed on nutrient and non-nutrient agar. (From Russell and Bonner, 1960.)

nutrient agar containing 1 per cent dextrose and 1 per cent peptone and phosphate buffer. As can be seen from Fig. 13c, the per cent germination on the nutrient agar was significantly higher. Recently Ceccarini (personal communication) has shown that if the dextrose is omitted, the peptone alone will produce the same effect.

Of greater interest is the observation that, as in many fungi,[2] the per cent germination decreases with increased spore density (Fig. 13a). In fact when the spore density is plotted against the per cent germination (Fig. 13b) an inverse relation is clearly evident.

One possible interpretation of these experiments is that the spores give off an inhibiting substance. This has been confirmed by Ceccarini (1966), who washed the spores and found that he could then achieve very high per cent germination with concentrated spores; the inhibitor had evidently been removed. Furthermore, spores placed in the wash water were inhibited in their germination.[3] It will be of considerable interest to discover the nature of this inhibitor and find what its effects are upon the spore.

We know very little about the biochemical changes that take place within the spore, although a few studies have been initiated. Clegg and Filosa (1961) showed that 7 per cent of the dry weight of the spore consists of the carbohydrate trehalose, and they suggest that this is the energy source for germination, which is the case in *Neurospora*, as has been shown by A. S. Sussman and his co-workers,[4] and in the resistant stages of a number of animals.[5] Furthermore

[2] See V. W. Cochrane, *Physiology of Fungi*. John Wiley and Sons, New York, 1958.
[3] It should be added that if the spores of two species are placed together with a minimum of water, one species will inhibit the other. The hierarchy of inhibition is D. *purpureum* > D. *discoideum* > D. *mucoroides* (Snyder and Ceccarini, 1966).
[4] A. S. Sussman, *Quart. Rev. Biol. 36* (1961): 102-116.
[5] See Clegg and Filosa (1961) for references.

in *Neurospora* after germination trehalose can no longer be detected. Recently Ceccarini and Filosa (1965) and Ceccarini (1966) have confirmed this finding and have shown that the enzyme trehalase is not present in the spores but shows high activity in the amoebae after germination. From the work on other forms it is appreciated that many additional chemical steps must take place in germination, for even a single developmental event such as germination involves a complex series of biochemical reactions.

One of the best pieces of evidence that the growth phase may be separated from other phases of the life cycle is that amoebae produced in the laboratory by germination of spores, if they are sufficiently numerous and concentrated, will aggregate and form a second generation of spores and fruiting bodies even in the total absence of food. This is all done on non-nutrient agar without any addition of outside energy. We have even been able to concentrate these second generation spores and make a third generation without food. Now that Ceccarini (1966) has shown that it is possible to remove the germination inhibitor by washing the spores, it may be possible to pursue this technique even further. It should be noted that one obvious consequence is that the cells (and the spores) become progressively smaller as starvation proceeds; there must be a definite limit to the number of generations which can be produced in this way.

2. Growth on solid media

Most of the experimental work on the cellular slime molds has involved solid media. This is true of all the early work and of the pioneer work of Raper (1937), who showed that the cellular slime molds would grow on a variety of different bacterial associates. In recent years two species of bacteria have been found to be especially effective: *Escherichia coli* and *Aerobactor aerogenes*. As Raper showed, many others are possible, but none are more effective than either of these.

There are two basic ways in which solid media have been used. In the first method, which has been developed and studied in detail by Raper (1937, 1939, and 1951), the nutrients in the medium are prepared so that both the bacterium and the slime mold will grow simultaneously. The other method, first used by Singh (1946), consists of mechanically placing on the agar surface bacteria which have been grown elsewhere and inoculating them with the spores of the slime mold.

Before going into these two methods, a few practical hints may be helpful to those who are working with slime molds for the first time.[6] It is possible to obtain pure spores, without the associated bacterium, by merely touching an aerial spore head with an inoculating loop. It is important to use fresh bacteria as a food source because, if the inoculation is made by scooping up both the spores and the bacteria of an old culture, the bacteria will often mutate and produce colonies that are gummy and inaccessible to the amoebae. These will therefore persist by selection, and soon one will have nothing but mutant bacteria that are totally inedible by the amoebae.

Also the slime mold may mutate or for other reasons it may be desirable to clone them, that is, obtain individuals from a single spore. This can be easily done, and a convenient procedure is described by Filosa (1962).[7]

The optimum temperature for growth is in the range 21-25° C. Growth will occur at higher temperatures, especially in some strains, but for general laboratory procedure it is

---

[6] The beginner may find the directions given by Lonert (1965) helpful. As he indicates, the General Biological Supply House, Inc. (8200 S. Hoyne Avenue, Chicago, Ill., 60620) now sells a complete kit, including medium, instructions, and cultures of *D. discoideum* and *E. coli* (61V15 *Dictyostelium discoideum* Culture Set).

[7] See page 167 for a general discussion of cloning.

not considered desirable to exceed 23° C. No abnormality appears at much lower temperatures (10° to 12° C) but it takes an inordinately long time for growth to occur.

The best source of information on methods of growing the cellular slime molds on a nutrient medium that permits the bacterium to grow and therefore the mold, is a comprehensive paper by Raper (1951). No attempt will be made to reproduce here the wealth of detail in that paper; I shall merely mention two media, one for sparse growth and one for heavy growth.

For sparse growth, Raper (1951) is especially partial to his hay infusion agar, and indeed this can be highly recommended. Somewhat easier to prepare, and very effective is another medium he suggests:

lactose	1 (or 0.5) g
peptone	1 (or 0.5) g
agar	20 g
distilled H_2O	1000 ml

All species of *Dictyostelium* and *Polysphondylium,* even the most delicate, will grow on this medium.

If a large quantity of one of the bigger species is needed (*D. mucoroides, D. purpureum, D. discoideum, P. violaceum*), we find the following medium useful:

peptone	10 g
dextrose	10 g
$Na_2HPO_4 \cdot 12\ H_2O$	0.96 g
KH_2PO_4	1.45 g
agar	20 g
distilled H_2O	1000 ml

This is convenient for growing stock cultures of these hardy species.

For biochemical work it is desirable that all the amoebae

be as closely as possible at the same stage of development, but in any ordinary culture dish, unless special care is taken, the beginning of aggregation will be spread over a long period of time, as will migration and culmination; there is no synchrony in the various stages of development. This means that if one harvests the amoebae after a certain number of hours, there is little assurance that one is harvesting amoebae from one stage of development. The first to examine this problem carefully was Takeuchi (1960), who tried a variety of different nutrients and found that the following rather weak nutrient gave the best results for *D. mucoroides* when 0.2 ml of a mixed spore and *E. coli* suspension was evenly spread over the surface with a sterile glass rod:

lactose	0.4 g
peptone	0.4 g
agar	20 g
distilled H_2O	1000 ml

Other procedures have been devised for obtaining synchrony in later stages of development, but these involve removal of the vegetative cells from nutrient-rich plates, washing them by centrifugation, and replating them on non-nutrient agar (Bonner, 1947; White and Sussman, 1961). Here, however, we are concerned with the growth phase, and we will shortly return to the matter of synchrony in this phase when we consider the liquid media, which have provided the best means of obtaining cells at a similar physiological age.

In some of the early growth studies, especially those of Raper (1939), the environmental conditions for optimal growth were examined with great care, but, while this approach was of considerable practical value, its value for determining the conditions of growth of the amoebae themselves was limited. The point is that the media had to permit both the bacteria and the amoebae to grow; the optimum conditions were those that were ideal for this association and not just for the amoebae alone. To give a specific illustration, Raper (1939) found that there was a sharp pH optimum for the two-membered culture, while Singh (1946, 1947a) showed that if the amoebae are somehow separated from the bacteria they will develop over a much larger pH range.

The method devised by Singh (1946) was to grow the bacteria separately and spread them with an inoculation loop on non-nutrient agar. This pre-grown bacteria could then be inoculated with the spores of a slime mold which would consume the static bacterial colony. Singh (1947a) used this method effectively as a means of isolating cellular slime molds from the soil, but anyone wishing to do this should also consult other sources for isolation methods, especially Raper (1951), Cavender and Raper (1965a,b,c), and L. S. Olive and Stoianovitch (1960).

If dense cultures and large pseudoplasmodia are to be obtained, it is convenient to make a sizable heap of the bacterial cells. The great advantage of such a system is that it will work with both the large, hardy species and the smaller delicate ones. We use this method routinely and find that the *E. coli* can be grown on either a solid medium (e.g. 1% dextrose, 1% peptone, 2% agar) or a liquid nutrient broth and then be washed by centrifugation; even a commercial frozen *E. coli* may be used.

It is also a most satisfactory method for obtaining thin, sparse cultures and has been extensively used by Shaffer. His (unpublished) modification of Singh's method simply consists of taking one small loopful of *E. coli* and painting it evenly with a smooth inoculation loop all over the non-nutrient agar surface of a petri dish. The spores of the slime mold can be spread on the surface, but they are better placed in one spot so that a ring of feeding and dividing cells expands outward, followed by a zone of aggregation and, within that, a zone of migration and culmination. This is by far the easiest method of demonstrating the stages of the life cycle to a class.

3. Growth in liquid media

Amoebae have been grown in liquid culture in the same two ways just described for solid media: (1) with nutrients, so that both the bacteria and the slime mold grow simultaneously, and (2) without nutrients, using pre-grown bacteria which are washed and suspended with some slime mold cells.

The method with nutrients has been tried in various laboratories with uneven success, but Sussman (1961a) has devised a dependable and reliable procedure. His medium consists of:

yeast extract	0.5 g
peptone	5 g
dextrose	5 g
KH_2PO_4	2.25 g
$K_2HPO_4 \cdot 12 \ H_2O$	1.5 g
$MgSO_4 \cdot 7 \ H_2O$	0.5 g
distilled H_2O	1000 ml

Sussman recommends 18 ml of this medium in 125-ml Erlenmeyer flasks, and we have also used large flasks successfully, provided the amount of medium is small in proportion to the flask size. After sterilization the flasks can be inoculated with either *Aerobacter aerogenes* or *Eschericha coli* (strain B/r— see below) and placed on a shaker at 22° C for 48 hours. This procedure is effective with *D. discoideum, D. mucoroides, D. purpureum,* but for some reason we have not been successful with *P. violaceum,* nor with any of the small, delicate species.

Although Sussman (1956b) briefly mentions that the

second method is possible, Gerisch (1959, 1960) was the first to describe in detail the method which involves growing the bacteria and then subsequently adding the slime molds to this source of food. He examined various strains of *E. coli* and found that while some formed ropes and were relatively inaccessible to the amoebae in liquid culture, some (strain B/r) remained edible in agitated culture. These he grew in a nutrient broth, washed them by centrifugation, and suspended them in a $M/60$ Sorensen's buffer at pH 6, at a concentration of 1×10^{10} cells/ml. The spores or the amoebae of the slime mold may then be added and growth will proceed rapidly.

If this procedure is to be used, the reader is urged to consult Hohl and Raper (1963a), who give clear directions and have made some additional modifications that are helpful and convenient. It should be added that one can now buy frozen *E. coli* that will serve as the source of bacteria to be added to the buffer.

Liquid culture with pre-grown bacteria is undoubtedly the best method we have for obtaining amoebae all at the same physiological age. Already, Gerisch (1960, 1962b) has made use of this important advantage, which comes from the fact that the medium is vigorously stirred or shaken and there can therefore be no pockets or areas that are further advanced or retarded than others, such as so easily occurs on a rigid agar surface.

Some observations made by Allderdice (1965) are of interest and should be pursued further. He discovered that if he made a series of flasks containing bacteria and Sorensen's buffer and stored them in the refrigerator, the older the flask, the shorter the lag period before logarithmic growth of the amoebae occurred. He was able to show that the older bacterial cells give off a substance which in some way makes them more immediately appetizing to the amoebae. Once growth begins, the growth rate is the same, independent of the age of the bacteria. The factor affects only the duration of the lag phase.

It has been known for some time (Potts, 1902; Raper, 1937) that cellular slime molds will grow on dead bacteria, but this new method of Gerisch makes it possible to study the phenomenon quantitatively, as has been done by Hohl and Raper (1963a). One of the interesting discoveries they made was that if the bacteria are autoclaved, the longer the heat treatment, the slower the growth of the amoebae feeding upon them. This is true (to varying degrees), for most strains, although they did find one strain of *Polysphondylium pallidum* (WS–320) that grew equally well regardless of how long the bacteria had been autoclaved. In an independent study, Gezelius (1962) confirmed the fact that *D. discoideum* grew more slowly on dead bacteria, but she could increase the growth rate by supplementing the medium with glucose, amino acids, and vitamins, and by lowering the initial pH.

4. Growth in cell-free media

The desirability of obtaining cell-free or axenic media or, even better, a medium in which all the chemical constituents are known, is obvious. Not only would all biochemical investigations be greatly enhanced, but there would be a possibility of obtaining nutritionally deficient mutants for genetic studies. The fact that cellular slime molds have to be grown on bacteria in a two-membered culture is a serious disadvantage.

The first reported attempt at cell-free culture was that of Bradley and Sussman (1952), but their procedure has never proved practical or repeatable; it has not been used (despite attempts to do so) by any workers since its original description. The matter was re-examined by Hohl and Raper

(1963b,c) in an exhaustive and lengthy study, and they were able, for the first time, to grow a cellular slime mold on a cell-free medium. Their approach was to use the methods of animal tissue culture and to screen a large number of different species and strains of slime molds to find the best combination.

In a medium containing embryo extract they were able to grow some strains (especially WS–320) of *P. pallidum* (Hohl and Raper, 1963b).[8] Using tryptose, serum albumen, and some inorganic salts, they grew two strains, but growth was significantly increased if vitamins, amino acids, dextrose, and trace elements were added (their complete medium: Hohl and Raper, 1963c). Finally they obtained good growth—although the growth rate was reduced—using the Hohl and Raper strain WS–320 of *P. pallidum.* Allen, Hutner, Goldstone, Lee, and Sussman (1963) have developed a defined medium for which they report moderate growth; they suggest that certain fatty acids are required.

The real test for these methods is whether or not they are sufficiently practical and reproducible to be used in experimental work. Recently Francis (1965) has used the Hohl and Raper method with complete success in a study of aggregation of *P. pallidum,* and it may be hoped that with continued work, the medium will be further simplified and made easy for all who wish to grow the organisms without bacteria.

5. Summary

Perhaps the best way to summarize the various growth methods described is to present a table of all the data in the literature on growth rates, in terms of generation times (Table 1). It can be seen at a glance that the question of

[8] The ability of Hohl and Raper's strain WS–320 of *P. pallidum* to grow on a cell free medium has been confirmed by Sussman (1963).

TABLE I

Summary of the values of the generation time for cellular slime molds grown on different media.

Species	Strain	Medium	Temperature in °C	Generation time in hours	Reference
D. discoideum	NC-4	Agar; 1% peptone + dextrose	22°	3.2	Sussman (1956a)
	H-1	Liquid; with nutrients and growing bacteria	22°	3.0	Sussman (1961a)
	RA	Liquid; with nutrients and growing bacteria	22°	3.6	Sussman (1961a)
	V-12	Liquid; with dead bacteria + nutrient supplement	25°	3.3	Gezelius (1962)
	V-12	Liquid; bacteria—suspension—Gerisch (1959)	23°	3.3	Gerisch (1959, 1960)
	V-12	Liquid; modified Gerisch method	25°	2.9	Hohl + Raper (1963a)
	NC-4	Liquid; modified Gerisch method	25°	2.6	Hohl + Raper (1963a)
D. mucoroides	11	Agar; 0.04% lactose + peptone	21.5°	2.8	Takeuchi (1960)
D. purpureum	WS-321	Liquid; modified Gerisch method	25°	2.4	Hohl + Raper (1963a)
P. violaceum	P-6	Liquid; modified Gerisch method	25°	2.4	Hohl + Raper (1963a)
P. pallidum	WS-320	Liquid; modified Gerisch method	25°	2.6	Hohl + Raper (1963a)
	WS-320	Liquid; with dead bacteria	25°	3.5	Hohl + Raper (1963b)
	WS-320	Cell-free medium; with embryo extract	25°	5.0-6.0	Hohl + Raper (1963b)
	Pan-17	Cell-free medium;	25°	4.5	Hohl + Raper (1963c)

whether the medium is liquid or solid, or the question of whether nutrients are present in the medium so that the bacteria can grow along with the amoebae or whether the amoebae are fed pre-grown bacteria, does not seriously affect the rate of growth. There is some slight reduction of the rate with dead bacteria, or with the cell-free media, but it is hoped that in the latter case this discrepancy may ultimately be eliminated.

This examination of the growth phase of the cellular slime molds has been limited more or less to a discussion of techniques; the biochemistry of growth synthesis is still to be elucidated. But inevitably one comes back to the fact that this is a phase of the life cycle that can be completely eliminated; it is not an essential step in the causal sequence. The amoebae from germinating spores need not be fed at all, yet they will still aggregate and form fruiting bodies. In the long run growth is obviously necessary, but it is not required for each cycle. The growth phase can be short or long, depending upon the food supply, but its duration has no appreciable effect on the morphogenesis of the sorocarps.

IV. Morphogenetic Movements

WHILE it is easy to separate growth from the other developmental processes in the cellular slime molds, it is exceedingly difficult to untangle morphogenetic movements from differentiation. We shall, nevertheless, do so in what follows for the simple purpose of making the discussion manageable and clear. The difficulty is that so often the two occur hand in hand, and to treat them separately appears artificial, but some form of systemization is essential unless one is willing to talk about everything at once.

Here we will consider under morphogenetic movements two principal topics. One is the mechanism of movement of the cells and the cell masses, and the other is the question of their orientation. All matters which involve the change of state within cells as they progress from one stage to another, and the external and internal factors which affect the changes, will be discussed as aspects of differentiation.

1. The mechanism of movement.

It is obvious, since the cellular slime molds are composed of amoebae, that the basic mechanism of movement is amoeboid. Unfortunately this tells us little, for despite years of intensive research we still are uncertain as to the mechanism of movement for any one type of amoeba, and there is a growing view that not all amoebae employ identical means of propulsion. It is not my intention here to indulge in a large series of hypotheses or in polemics, but I should like to present what facts are known and from these make some simple tentative hypotheses.

The most difficult question is that of the mechanism of movement of the amoebae themselves; very little is definitely settled about the whole problem of amoeboid movement. Shaffer (1964a, 1965a,b) has made careful observations on isolated amoebae and finds a number of interesting phenomena, all of which lead him to suspect that the motive force for the movement of these amoebae lies in the thin surface layer of the cells. The evidence for this is based primarily on the fact that although he observes within the cells very rapid saltatory movements of vacuoles and particles that move right up to the cell surface and away again, these movements appear to be quite unrelated to cell movements; they are far more rapid, and their direction is haphazard and bears no relation to that of the whole cell. Furthermore, since the vacuoles and particles move, without any decrease in speed, right up to the cell membrane (as seen in a light microscope), there is no evidence that these amoebae form any extensive area of plasma gel, as some of the larger amoebae do.

As is well known, amoebae in general are polarized; that is, they have a "head" and a "tail" region, and the slime mold amoebae may maintain such polarity for extended periods of time. At the anterior ends they frequently produce their pseudopods, which, if the amoebae are in chains, connect one cell to the next. After the pseudopods are formed they often become thin, and these changes in morphology could, according to Shaffer, be explained by assuming that the new surface is made at the tip and that surface is reabsorbed at the base of the thin pseudopod. He makes the further hypothesis that the same process is occurring in the whole cell; that is, the anterior end in general is characterized as a region of new surface, and the posterior end is a region of the reabsorption of surface. It is important to keep in mind the fact that this is an hypothesis, even though it is exceptionally helpful in explaining movement in the cell masses.

The main fact which supports the hypothesis is that the cell surface along the sides of the cell or of a thin pseudopod is rigid and immobile.

In aggregation the amoebae form chains, and here again Shaffer (1964a) has made a significant observation. The cells tend to adhere to one another at specific sites; the anterior end of one will attach to the posterior end of another. The best evidence for this phenomena comes from cells which enter a stream from the side (Fig. 14). They do not remain fixed to the middle of a stream cell but move forward or backward and adhere to the posterior end of one of the cells.

Far less is known about the mechanism of movement dur-

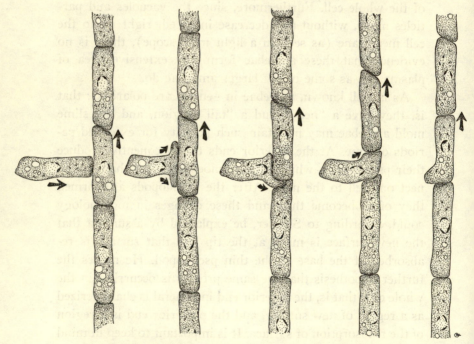

Fig. 14. Drawing of an amoeba entering an aggregating stream. Note that ultimately it achieves an end-to-end attachment. (Redrawn from B. M. Shaffer, 1964a.)

ing migration than is known for the aggregation stage, mainly because during migration the cells are bunched together and it is more difficult to observe them individually. Most of the work on migration has been performed on *Dictyostelium discoideum*, but there is no reason to believe that the problems are not the same for those species that manufacture stalk during migration. Since a discussion of the movement during culmination will follow, the special problem of the movement of the cells actively forming a stalk will be considered then.

The most obvious kind of factual information that can be gathered for migration is on the speed of movement of the mass. This has been done in some detail for *D. discoideum*, and it was found that the larger the migrating pseudoplasmodium, the faster it moves (Bonner, Koontz, and Paton, 1953; Francis, 1959, 1962). It has even been possible to follow individual pseudoplasmodia that become smaller with continued migration, and one finds that the rate of movement drops correspondingly. At 20°C the range of rates extends from 0.3 to 2.0 mm/hour.[1]

It is possible to compare the speed of movement of migrating masses with those of individual amoebae; this has been done in detail by Samuel (1961). It can be seen, from Fig. 15 that the rates at the various stages are all roughly within the same range.

A perplexing number of hypotheses have been advanced to explain the simple fact that large slugs (and larger culminating pseudoplasmodia) move faster than new small ones (Bonner and Eldredge, 1945; Bonner, Koontz, and Paton, 1953; Francis, 1962; Shaffer, 1962, 1964a, 1965c; Bonner, Kelso, and Gillmor, 1966). The only point of general agreement is that this fact suggests that all the cells

[1] Raper (1956b) finds the rate of the migrating pseudoplasmodium of *D. polycephalum* about 0.5 mm/hour, which is consistent with *D. discoideum*.

(not just a surface layer of cells) contribute to the movement, although even this is pure conjecture; there is other evidence which supports this contention more effectively. Namely, if the migrating slug is allowed to crawl onto a coverslip, and the coverslip is then inverted in a small, moist chamber, it is possible to observe the cells with high powers of the microscope for some depth. Such observations could be profitably extended and a detailed analysis made, yet even a short glance will show that the cells are all actively pseudo-

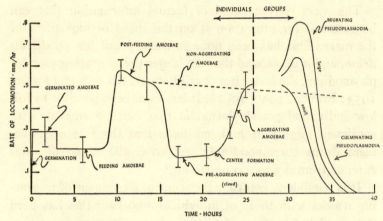

Fig. 15. Changes in the rate of locomotion of amoebae throughout the life cycle of *D. mucoroides*. The vertical lines indicate the standard deviation (each point is an average of 60-120 determinations). After aggregation it is no longer possible to follow the rate of the individual cells, and the rates of migrating pseudoplasmodia are included for comparison. (From Samuel, 1961.)

podial and that they do not move at exactly the same rate; their relative positions will vary to some extent over the course of a few minutes.

The fact that all the cells do not appear to be moving at the same rate has been examined in another way. When Raper (1940b) made his original grafts of white and red fractions of migrating pseudoplasmodia (stained by growing the cells on the red bacterium *Serratia marcescens*), he showed that the division line stays constant for a number of hours. Since then, we have found two exceptions to this general rule.

The first is at the end of aggregation. We have made grafts during the aggregation stage: a colorless aggregation center is placed in front of an aggregating stream in which the amoebae have been vitally stained with Nile blue sulphate (Bonner and Adams, 1958). Theoretically this should produce a migrating slug in which the anterior half is colorless and the posterior half blue (or vice versa, if a blue center is grafted onto colorless amoebae), for such would be the case if the graft were made during the migration stage. The fact is, however, that the slug will be uniformly blue, indicating that between aggregation and migration there is a critical period in which there is a violent redistribution of cells. There is no known counterpart at a later stage; it is only at the end of aggregation that one has this great mixing of the cells within the mass.

The second exception to the notion that during migration the cells keep a fixed position is the observation that the division line does not remain completely sharp; there are a few cells which wander past it (Bonner, 1952, 1957). If a colorless tip is grafted onto a hind end stained with Nile blue sulphate, individual blue cells can be seen wandering forward in the colorless tip. If the reverse graft is made, individual blue cells can be seen to fall backward, indicating that throughout migration there are a few cells that are especially fast and are moving forward, and a few that are especially slow and lag behind. It is even possible to measure the speed of these cells; in a large pseudoplasmodium it takes approximately six hours for a fast cell to pass from the mid-region of a migrating sausage to the tip. It must be understood here that all we observe is a relative change of position of the cells. We do not have any direct evidence that these cells are shifted because of different intrinsic rates of motion. There are other possible explanations, such as selective adhesive states, and therefore by "fast" and "slow" we can only mean relative position within the pseudo-plasmodium.

These vital-stain experiments have been done entirely on *D. discoideum*, but there is evidence that there are at least slow-moving cells in *D. mucoroides* (Bonner, 1957). These accumulate at the posterior end and take on the staining properties of the pre-stalk cells; earlier in this book I have referred to them as the rear-guard cells. In a particular strain *D. mucoroides* it is possible to measure the increase in the size of the rear-guard zone as migration proceeds. For a long period there is a slow accumulation of cells in this posterior zone, but after 5 cm of migration the rate suddenly rises sharply. In *D. discoideum*, where the rear-guard cells apparently form the basal disc, there is no such accumulation, and the zone stays approximately the same size. The reason for this is that the cells are constantly left behind and lost in the slime track as mentioned earlier. Therefore the size of the basal disc is the same for those individuals that have migrated hardly at all and those that have migrated great distances.

The fact that all the cells do not have the same potential of movement was also ascertained by another experiment (Bonner, 1952). If some vitally stained cells from the anterior end of *D. discoideum* are grafted onto the posterior end of a colorless migrating slug, then in a matter of two or three hours this colored segment moves up to the anterior position corresponding to its original location. While moving, the colored mass remains discrete, and as the whole sausage migrates, the colored cells move faster in a sharp band which finally reaches the forward position. If a group of colored posterior cells is transplanted into the anterior end of a sausage, they slowly lag and eventually assume a posterior position. If anterior pieces are placed in anterior positions and posterior pieces in posterior positions, their location remains fixed even after prolonged migration.

Another instance of a change in the relative positions of the groups of cells is an experiment in which the center of *D. discoideum* is placed among the aggregating streams of a particular strain of *D. mucoroides* (Bonner and Adams, 1958). Presumably this should produce a migrating mass with the anterior end *D. discoideum* and the posterior end *D. mucoroides*, but apparently the *D. mucoroides* cells are faster than the *D. discoideum*, for they pass through the latter and take over the anterior position. If the reverse graft is made, the anteriorly placed *D. mucoroides* remain in position and there is no change.

The above reversal in position may well take place during the period of violent cell redistribution between the aggregation and migration stages. This, however, cannot be the case when the vital dye grafts are made during the migration stage. It may be that there is in any population of cells a range of cell velocities. After the end of aggregation these cells soon sort themselves out, with the relatively fast ones in front and the relatively slow ones behind. As migration proceeds, a few stragglers continue to shift their positions, and some cells may even change into either fast or slow cells and then shift their positions. The implications of these differences in rate of movement, as far as differentiation is concerned, will be discussed in the next chapter, but here, in a consideration of the mechanism of migration movement, the differences underline the fact that each amoeba is relatively independent in its movement, and that to some extent the whole migrating sausage is a mass of such independent cells bound by their stickiness and the slime sheath.

· 99 ·

To turn now to another related problem, there is a great deal of evidence to suggest that traction occurs between the slime sheath and the amoebae rather than between the sheath and the environment. The sheath is stationary and is laid out as a carpet upon which the amoebae may tread. Markers placed on the sheath do not move as the cell mass pulls out from under them; also the whole migrating mass will move at the same rate, whether it is touching the substratum over most of its surface or whether it is partially pointed up into the air. This latter condition may be so exaggerated that merely the hind end of the slug is touching the substratum, a condition especially striking in the case of the long, string-like pseudoplasmodia of *D. polycephalum*. Also, both *D. discoideum* and *D. polycephalum* will migrate for great distances across the hyphae of some common mold without touching the agar surface, merely touching one hypha after another as though they were passing through the branches of some minute tree.

Recently Shaffer (1965c) has obtained much more information on this point by extensive studies with markers on the sheath surface. Of special interest is the fact that if a slug is suspended, by its base, from the ceiling of a culture dish, the markers will move along the slug toward the base even though the slug is not moving. In fact there will be a great accumulation of sheath material at the base if this condition persists. This provides striking evidence that the traction is indeed between the sheath and the amoebae inside it.

Since the amoebae secrete the sheath and then move along what they have made, it is not surprising to find the sheath thinnest at the tip. Francis (1962) has been able to show that the sheath is fluid at the tip, but that further back on the slug it is no longer plastic and exhibits the properties of a solid. Presumably, the material must congeal in some way after it is exuded.

There are, then, three main points of information upon which we may build a hypothetical view of the mechanism of movement: (1) the slime sheath is stationary; (2) all the cells within the mass are actively moving; and (3) the cells inside the mass move at approximately the same rate as cells outside the mass. This suggests that movement within the migrating mass is basically no different from that of an aggregation stream. All the cells are moving forward, but as a migration slug they are in so thick a mass that the question soon arises as to how the cells in the interior of the mass obtain traction. The case is clear enough for those at the edge, because they are adjacent to the slime sheath, but does the sheath material extend inward like a net to give traction to all the cells? Unfortunately we do not know the answer to this question, for the sheath is so extremely thin that it is impossible to follow it internally, and whether all the cells or just the external ones secrete it cannot be determined as yet. However, an internal slime net may not be necessary to explain how traction is provided for the internal amoebae. When amoebae move, much of their external wall expanse is rigid, and the central channel of protoplasm which pours out to the tip is relatively fluid. If the walls are rigid (at least for short periods of time) and the cells are sticky and adhere to one another, then they will provide one another with temporary traction. Each amoeba lays down its own carpet in the form of thin, solid wall; the walls in turn stick to one another and ultimately to the slime sheath. It is true that these adhering walls are constantly torn down and built anew perhaps as Shaffer suggests, but at any one instant there is a firm framework or skeleton of rigid wall that, like a three-dimensional net, penetrates the whole migrating cell mass.

There is one further puzzling observation which concerns the orientation of the cells, and this observation applies to both migrating and culminating pseudoplasmodia. One of the striking evidences of order in the cell masses in general is that all the cells are moving in the same direction. This polar movement is evident from the beginning of aggregation onward, and presently we will discuss the possible causes of this orientation. In aggregating amoebae the long axes of the cells lie in the direction of the movement—a most reasonable configuration since we would expect any amoeba to be longest in the direction in which it is going. But for some reason that is not understood, the amoebae within the migrating masses are either isodiametric or spindle shaped at right angles to the direction of movement (Fig. 16). The latter configuration is present in the tip of the migrating slug of *D. discoideum*, especially when the tip is thin and tapered.

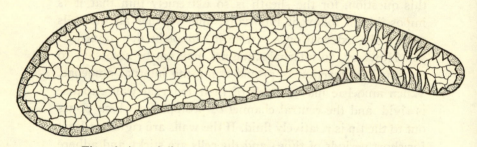

Fig. 16. A semi-diagrammatic view of a section of a migrating pseudoplasmodium of *D. discoideum* showing the typical transverse orientation of the cells in the narrow anterior portion.

In culminating cell masses it is also present at the tip in the cells that surround the stalk (Fig. 8) and occasionally in the rear-guard cells, posterior to the presumptive spore cells. It is especially evident in the anterior cells that are actively secreting the cellulose to form the stalk sheath at their centripetal ends (Bonner, Chiquoine, and Kolderie, 1955).

A variety of hypotheses have been made to account for the puzzling fact that cells are elongate at right angles to their direction of movement (Raper and Fennell, 1952; Bonner, 1959b; Shaffer, 1962), but in retrospect they all involve so much conjecture that they seem unsatisfactory. We do not even know the answer to the main question: Are these cells actually moving (Shaffer, 1962) or are they being pushed or squashed (Raper and Fennell, 1952)? It is time for some new facts.

In culmination there are two other observations of significance to the problem of movement. It has been shown, in *D. discoideum*, that during culmination there is a steady decrease in rate of upward movement as maturity is reached (Bonner and Eldredge, 1945). There is no reflection of the fact that the encapsulation of the majority of the spores occurs quite rapidly, within an hour or less, relatively early in the rising process (Bonner, 1944), and this has led to the suggestion that the pre-stalk cells (and the rear-guard cells below the pre-spore cells) are largely responsible for the upward thrust. In those forms with a cellular stalk there is also the further factor of the expansion of the pre-stalk cells by vacuolization once they have entered the stalk. As Raper and Fennell (1952) suggest, this could contribute to the lengthening of the rising cell mass, but Raper also showed (1956a, Raper and Quinlan, 1958) that this could not apply in the case of *Acytostelium* with its acellular stalk. The whole question of the source of the motive force for the movement involved in migration and culmination is still a matter of conjecture.

2. The control of the movement

Morphogenetic movement always consists of two features: the movement itself, and its control resulting in the emergence of a consistent shape. Movement without coordination would be chaos. In this section we will consider the various methods of control, including the minor question of rate of the cell movement and the major question of the orientation of the cell movement. It is this latter governing of direction that is so essential for the production of a consistent form.

The first detailed and comprehensive study of the rate of cell movement is that of Samuel (1961). He showed that during the life cycle there are periods of different cell speeds, which show up as a consistent pattern from generation to generation (Fig. 15). This led him to conclude that there must be substances given off by the cells which control the rate of movement of themselves and their neighbors; the rate changes must be caused by rate-limiting substances given off by the cells themselves. We shall have a good deal more to say about "rate substances" later, but first it is important to give some background on the problem of the mechanism of orientation in aggregation. This is one of the oldest problems in the study of the cellular slime molds and remains one of the most interesting. Aggregation is so conspicuous a morphogenetic movement that it would seem especially amenable to experimental analysis. To be able to understand fully the control of any morphogenetic movement would be a significant advance in developmental biology.

Before beginning with the details, let me make a few generalities about this type of problem as it has emerged over the last two decades. In any analysis of development one begins with a simple description, and this is followed by an experimental analysis to determine what are the principal controlling factors in the process. But if this biological approach alone is used, with further analysis one soon becomes lost in a maze of unexplained effects. One goes through a cycle in which the first biological experiments are clarifying and show the general nature of the problem, but in which the qualifying, subsidiary observations and effects soon become so complex that it is no longer possible to have a clear picture. It is at this point that a deeper chemical insight into the problem is essential, and when the substances involved are known, it is possible slowly to piece together the chemical information and the biological phenomena. Advances in various areas of hormonal work, both animal and plant, have undergone these cycles of clarification and obfuscation. In the long run we come closer to the truth, although there are moments when one seems to be going in the opposite direction. This certainly has been illustrated in some of the advances in work on aggregation, and we still have many more crests and hollows to ride out before we come to any final solutions of these intriguing problems.

Both Olive (1902) and Potts (1902) independently made the suggestion that the aggregating cells are oriented to the central collection points by chemotaxis. Neither provided any evidence to support the hypothesis, although Olive did attempt, without success, to orient the amoebae by creating gradients with solutions of sugar and malic acid. This experiment was stimulated by the work of Pfeffer[2], who had such striking success orienting the spermatozoa of bracken with these substances.

No further experimental work was attempted until 1942, when Runyon performed a notable experiment: If amoebae were put on two sides of a cellophane membrane, the aggregation streams on both sides were perfectly aligned, one above the other. In other words, the factor responsible for orientation could readily pass through the membrane. Runyon (1942) considered this further evidence for chemotaxis, but other possibilities remained, for a wide variety of agents other than small molecules could readily pass through cellophane. In view of this fact I attempted a number of experiments, examining the possibility that electric and other forces might be operating, but all save one provided negative results (Bonner, 1947). The successful experiment gave excellent evidence that chemotaxis was indeed involved. If the amoebae

[2] See Lord Rothschild, *Fertilization*. Methuen, London, 1956.

were allowed to aggregate on the bottom of a dish under a layer of water, and the water was moved gently over their surface (as though they were lying on the bottom of a brook), then the amoebae upstream of the center showed no orientation toward it and wandered about aimlessly, while the amoebae downstream were beautifully oriented toward the center and would move upstream from considerable distances to join it. This gave positive evidence of a free diffusing agent, and it was immediately evident that the agent was probably a chemical substance rather than heat, because it was known that the center of some species would not attract the amoebae of other species. This substance was given the generic name of "acrasin," which was intended to designate the chemotactic agent of any member of the Acrasiales.

The next step was to produce the substance *in vitro* and in this way have an absolute demonstration of its existence. I made numerous attempts to do this by taking agar on which a center presumably emitting acrasin had rested, and by taking freshly killed centers (killed in a wide variety of ways) and placing these non-living sources in among sensitive amoebae. In no case was there any evidence for chemotaxis. There were a number of reasons to suspect that acrasin either broke down rapidly or was volatile, and that since the agar blocks or dead centers were unable to produce new acrasin, what was there originally was soon lost and the gradient obliterated. Of course, even if there were acrasin present and it had a uniform distribution, there would be no chemotaxis, for the orientation of cells by a diffusing chemical can only be achieved if the chemical is in a gradient; there must be more molecules bombarding one end of a cell than the other, thus providing information about direction to the amoeba. Experiments were devised to see if aggregation could occur across a moist air barrier rather than the usual aqueous barrier, but there was no evidence for a volatile chemotactic agent from these experiments.

In 1950 Pfützner-Eckert, working in the laboratory of A. Kühn, reported that she was able to obtain orientation toward agar blocks that had been in contact with centers of *D. mucoroides* and then placed among sensitive amoebae. This is quite contrary to my experiments on *D. discoideum*, and Shaffer (1956b) reports that he was unable to repeat her experiments with *D. mucoroides*. The explanation of this enigma is unknown, and it is hoped that it will someday be forthcoming. Shaffer also reports many other types of unsuccessful experiments using agar blocks in different ways and killing active acrasin-secreting sources in a variety of manners.

In order to circumvent the possibility that in such experiments the acrasin gradient disappears, Shaffer (1953, 1956b) devised an ingenious experiment which provided the first unequivocal demonstration of the chemical nature of the aggregation phenomenon. Sensitive amoebae are sandwiched between a glass slide and a minute block of agar (2 to 3 mm square). This is placed within a moist chamber, and alongside on the glass slide are put a number of cell masses actively producing acrasin. A small drop of water is added beside each one of these acrasin sources. With a fine pipette or a glass rod a cell-free drop[3] is removed from the vicinity of one of the sources and placed on the side of the agar block so that a meniscus is formed. This same operation is repeated at short intervals, each time taking the drop from a fresh source. The amoebae in the glass-agar sandwich will, in five to ten minutes, show an orientation to the nearest edge of the block, and then proceed to move toward it.

Consider now the logic of this experiment. The amoebae are wedged under an agar block so there can be no convection, as would be the case were the amoebae merely placed under water. The meniscus at the edge provides a firm site

[3] This was checked by passing the drop through a millipore filter before using it (Shaffer, 1956b).

where the acrasin source remains fixed. The gradient is maintained by the repeated additions of fresh acrasin-water, for if the acrasin is destroyed and if diffusion tends to obliterate rapidly an original gradient, this will be counteracted by supplying the material at constant intervals over a period. It should be added that Shaffer has tried a variety of controls, using plain water and differently shaped blocks and adding the acrasin-water in different ways; these have all supported the interpretation that this is a valid and clear-cut *in vitro* demonstration of acrasin.

The next step was to examine its stability. By holding the acrasin-water in a pipette for various time intervals before applying it to the meniscus, Shaffer showed that the acrasin lost its potency in a matter of a few minutes and no longer was capable of orienting the amoebae. With fresh acrasin-water positive results were obtained if the intervals between application at the meniscus varied between ten seconds and two minutes. If, however, the acrasin-water was stored five minutes before application, it had either a weak effect or none at all, even if it was reapplied at short intervals. On the other hand, when these pipettes full of acrasin-water were quickly frozen with dry ice, they could be stored for long periods and, if quickly warmed up and used at short intervals, retained their original activity. Clearly, then, acrasin was for some reason unstable at room temperature, but in a frozen condition it retained its full potency.

In further analyzing the problem of acrasin stability, Shaffer (1956a, b) found that if he allowed the aggregation streams or acrasin sources to lie on one side of a dializing membrane, and if he then collected acrasin-water from the opposite side, the acrasin was stable for long periods at room temperature. The obvious inference is that the acrasin is normally destroyed by some substance of high molecular weight, presumably a protein enzyme. He also has shown that if cold methanol was poured over cultures in petri dishes and then vacuum-dried in the cold, and the residue eventually brought into aqueous solution, this preparation was both active and stable.

Following this work on stabilization, Sussman, Lee, and Kerr (1956) found a totally different method of obtaining stable acrasin. They immersed cultures in cold, dilute HCl at pH 3.5, and the resulting solution was shown to be stable and active, using Shaffer's test.

There are a number of reasons, as Shaffer (1956b, 1957a) emphasizes, why the destruction of acrasin might be of great importance in the aggregation process. It would effectively increase the relative gradient of acrasin, as well as lower its total concentration. The lowering of the total concentration is especially important where the amount of liquid medium surrounding the aggregate is small, as in aggregation taking place on a thin film of moisture on the surface of a glass slide. High concentrations of acrasin in such small quantities of water would soon produce a large "background noise," making the problem of detecting a gradient most difficult.

In a series of papers Shaffer (1956b et seq.) has described many further observations on the aggregation process and has speculated in a most stimulating manner as to their meaning. Let us here examine a few particularly significant contributions.

He showed that if sensitive amoebae, about to aggregate, are bathed by acrasin from a nearby source, they will become sticky and produce acrasin of their own. The acrasin itself initiates these reactions. It might be thought that this stage of integration is only reached when an amoeba enters a stream, but the action can occur at a distance. To ascertain this point Shaffer kept moving a center away from some separate amoebae that were approaching it, and finally the separate amoebae clumped together into a compact, sticky, acrasin-secreting stream that had no connection with the constantly dislodged center. Shaffer suggests that the acrasin actively transforms the cells into a new state, one in which there has been a change in the surface properties of the cells so that they will now adhere to one another.

It had been known before that once the cells do adhere, their individual movements affect one another. This was shown, for instance, in some experiments, already cited, in which a section of an aggregating stream was cut out, turned 180°, and reattached to the stump that was entering into the center (Bonner, 1950). The reversed piece was then literally pulled into the center as though it were a piece of treacle. The cells in the intact stump of the stream were actively moving toward the center and exerted a pull-tension on the reversed segment of stream, causing the cells within that segment to follow the tension. This is consistent with views of Weiss[4] that cells are affected by mechanical forces in their immediate environment, all of which means that besides chemotaxis, an important factor in orienting the cells to a common center is the mechanical one of pull-tensions. This is possible because the cells have become sticky and adhere to one another—a property which arises as the result of the action of acrasin on the cells. Therefore, even though aggregation must be thought of in terms of both chemotaxis and mechanical factors, acrasin is involved in both aspects and is, in this sense, the overseeing, controlling agent in aggregation.

But one important fact has been left unexplained. How do the cells within a stream become oriented so that they all move towards the center? This brings us again to the intriguing problem of polarity, and in a stream there would seem to be a rather rigid polarity of movement. All the cells are

[4] *J. Exptl. Zool. 68* (1934): 393-448; *100* (1945): 353-386.

going the same way and they appear mutually to pull at one another so that there can be no strays, no contrary individuals; since pull-tensions are most effective in orienting the cells, they will all go the way the majority goes.

That the direction of movement within a stream is mechanical is supported by a revealing set of experiments of Shaffer. He tested the acrasin emission of different portions of streams and centers by placing the streams and the centers in competition for some random amoebae. In this way he showed that the streams emitted just as much acrasin as the centers, and there was no evidence at all for an overall gradient of acrasin in the aggregate as a whole. As he points out, it is impossible, at the moment, to detect what small internal (and perhaps ephemeral) gradients there might be within the mass, but the evidence suggests that the cells in the stream work on a mechanical follow-the-leader principle, and that the direction must be imparted to them in the beginning. A final bit of evidence to support this concept is the formation of rings (Arndt, 1937; Raper, 1941; Bonner, 1950; Shaffer, 1957c). These frequently occur at the center of an aggregation pattern, and there is no apex, but a circle of cells adhering to one another and continuously moving around.

Originally it was assumed that during aggregation there was one large gradient over the whole aggregation pattern. This, as Shaffer points out, is hardly possible, and in its stead he has suggested an ingenious hypothetical explanation which assumes a series of directional impulses, one following the next. The center, which in this hypothesis is the pacemaker, emits some acrasin, and a gradient is set up in its immediate vicinity. The adjacent amoebae are oriented and eventually stimulated to produce acrasin. The orientation, however, occurs first, before the first surge of acrasin has been largely inactivated, and before these secondary amoebae produce a surge of acrasin of their own. This new surge produces a new gradient which now spreads to the amoebae beyond it and produces similar orientation and subsequent stimulation. Since the gradient is first in a position to orient the cells correctly toward the center, the fact that subsequently there is either no gradient or a reverse gradient, is insufficient, according to the hypothesis, to override the effect of the first gradient, which acted upon the cells while they were in a receptive period before they became "fatigued" and before they secreted their own acrasin. If the center emits acrasin rhythmically, then with each new impulse the peripheral amoebae will receive a new directional message; in fact by repeating the message, the center retains its key position as a pacemaker.

These impulses could be, as Shaffer points out, the rhythmic waves of fast inward motion revealed by the time-lapse motion pictures. These waves start in the center and radiate outward; they are waves of fast inward motion of the individual amoebae.

Once the amoebae are oriented towards a common point, they will coalesce into streams, for acrasin not only induces the cells to produce their own acrasin, but induces stickiness as well. The streams themselves actively produce acrasin, and therefore they can in turn attract separate amoebae in just the way a center is capable of doing. It is possible to see the rhythmic waves pass down a stream, and therefore the ability to relay the acrasin emission is presumably possible within a stream as well as between separated amoebae. It is for this reason that the importance of acrasin activity in orienting the amoebae within a stream is in some doubt, although the fact that mechanical factors play a significant role remains an excellent possibility.

The relay hypothesis of Shaffer has the merit of explaining two facts which would seem to be inconsistent with each other: Chemotaxis unquestionably occurs, but Shaffer was often unable to demonstrate any difference in the amount of acrasin secreted by centers and different parts of streams. These facts are satisfied by the hypothesis that there is no overall gradient, but a relay system of small ephemeral gradients. A further advantage of the idea is that it can explain the large aggregation expanses of oriented amoebae that occur before the final aggregation patterns in some species. It is a method that can operate at a great distance, provided there are scattered intermediate cells to carry the message outward.

The discussion of aggregation given thus far is a summary of what might be considered the standard views which have dominated the literature for the past two decades. Since Shaffer's work very little progress has been made, chiefly because we are now at the stage where it has become essential to know the chemical nature of acrasin and of any other substance which might be involved in the aggregation process. If we do not have this information, our only other recourse is to make increasingly elaborate hypotheses.

The first attempt to identify acrasin was that of Sussman, Lee, and Kerr (1956), who reported two fractions which had to be mixed in specific proportions to obtain activity. Later, Sussman, Sussman, and Fu (1958) showed there to be a third factor in the complex, but they were unable to identify the specific substances. Wright and Anderson (1958) noted that urine of pregnant women was active in the Shaffer test and proceeded to show that this activity lay in some of the steroid hormones in the urine. They then examined the slime molds for steroids and discovered that they synthesize stigmastenol. There is, among workers in the field, grave doubt as to whether or not this substance is acrasin. The doubt comes in part from the great difficulties in making the Shaffer test operate satisfactorily, and in part because for so many years the animal embryologists were fooled in thinking the primary inductor in amphibian gastrulation was a steroid, whereas it turned out that the steroid merely caused the cells to secrete the normal evocator substance. These qualms have been reinforced by the work of Hostak and Raper (1960), who showed, using another technique of Shaffer, that various alkaloids have some sort of acrasin-like effect on the aggregation of *Acytostelium.*

The difficulty has been that none of the assays developed so far are suitable for chemical analysis. An excellent new quantitative assay, devised by Francis (1965), involves the flowing of test solutions over cells aggregating under a layer of gently moving water. This test may be very useful in throwing new light on the aggregation mechanism, but, like the Shaffer test, it is technically too difficult and demanding for a chemical analysis.

In an attempt to find a simpler assay, Bonner, Kelso, and Gillmor (1966) prepared stable solutions of acrasin by previously known methods of Shaffer (1965a, b) and discovered that in the presence of these solutions the amoebae moved more rapidly. Furthermore, the relation was quantitative within a wide range; the higher the concentration, the greater the speed of movement. The question arises as to whether the preparation contains both an orientation factor (acrasin) and a separate, rate-increasing factor, or whether acrasin performs both functions. Despite considerable chemical purification we have been unable to separate the two components, the orientation being tested by the Shaffer method. This matter can only be settled when the chemical analysis is finished and we know the substance or substances involved. For the sake of convenience and accuracy, we will provisionally refer to the "rate substance" and acrasin (the orientation substance) separately and will hold in abeyance the question of whether or not they are one and the same.

Using this rate test in conjunction with the Shaffer test we examined various stages and species to find when and by what this substance(s) is produced. Surprisingly no species specificity was found, even between *D. discoideum* and *P. violaceum*. The substance(s) produced by one oriented and increased the rate of the other and vice versa. This was contrary to the surmises of Shaffer (1957a), who postulated separate acrasins for these species, although he did point out that it was possible to interpret his results on the basis of one substance if the assumption of differential response to it by the two species was made. Another surprise was the discovery that the substance(s) was produced in greater quantity hours before aggregation than during aggregation. The final and most astonishing result of all, was that *E. coli* was an even more potent source than the amoebae and that aggregating amoebae would stream towards heaps of *E. coli* (Fig. 17). Not only that, but these *E. coli* preparations had the power to make cells adhesive and clump in exactly the same fashion that was reported for acrasin by Shaffer (1957a), as we have previously described. It was also possible to show that the phenomenon of cell repulsion (originally described by Samuel, 1961) is totally unrelated to this rate and orientation substance(s) and is presumably mediated by another substance(s). [*See addendum on p. 124.*]

The idea that rate increase and orientation might go hand in hand is consistent with one old observation. As was pointed out previously, cells often move inward in waves or pulses of rapid movement that start in the center and radiate outward. In other words if the pulses are indeed periodic puffs of acrasin, these puffs simultaneously accelerate and orient the amoebae.

Another observation made from an old film would also fit in with the idea that rate of movement and orientation go together. When a center is removed during aggregation the

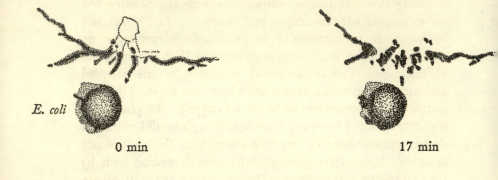

E. coli

0 min   17 min

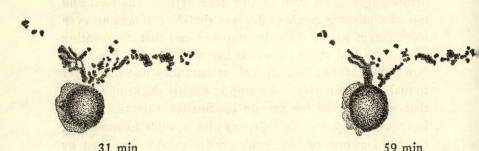

31 min   59 min

Fig. 17. The effect of a mass of *E. coli* on aggregating amoebae of *D. discoideum*. (From Bonner, Kelso, and Gillmor, 1966.)

streams break up, but they continue to pulse; the cells repel one another but do so in definite intervals of fast movement. If, as has always been assumed, the pulses are associated with acrasin secretion, then how is it possible that cell repulsion rather than aggregation occurs?

This paradox is eliminated if we consider aggregation to be characterized by a period of especially high orientation sensitivity to acrasin, so high in fact that it overrules the cell repulsion phenomenon. According to this hypothesis then, the removal of the center causes a momentary loss of orientation sensitivity on the part of the cells, not a loss in acrasin secretion, the existence of which is demonstrated by the pulses of fast movement.

This would also explain why the Shaffer test only works with amoebae that are in the process of aggregating, and not with cells just about to aggregate. One must have cells which are in this state of high sensitivity.

The important point is that if we assume all orientation and rate increase to be brought about by one substance, our traditional picture of the mechanism of aggregation must be radically altered. We can no longer think of the onset of aggregation as the moment when cells begin both to secrete acrasin and to become sensitive to it. Acrasin is always present, even in the food source, and is presumably a means by which the amoebae reach the food. Samuel (1961) discovered the fact that clumps of *E. coli* have a definite although weak ability to orient the vegetative amoebae, but the separate phenomenon of cell repulsion keeps the cells to some degree dispersed. After the food supply is depleted something occurs rather suddenly within the amoebae that produces a great increase in orientation sensitivity. Thus we see that a single chemotactic factor can be used in the first instance as a means of finding food and in the second as a means of aggregating and forming a communal, spore-bearing structure.

Along with cell repulsion, it is clear that center formation and center inhibition (which will be discussed in the next chapter) are processes distinct from the orientation-rate factors described above. Besides these separate problems there are a number of others that need thorough investigation. For instance, there is the whole matter of how gradients of acrasin arise. This may be related to the interesting observation of Samuel (1961), who showed that the amoebae had different rates of movement during different periods of their development, the period just prior to aggregation being one of especially slow movement (Fig. 15). Another matter that needs further investigation is the mechanism of species specificity during aggregation. Finally there is the old question of whether or not the substance(s) is performing some function in the cell masses, for it has been known for a long time to be present there (Bonner, 1949).

Thus far we have confined our remarks to orientation of the amoebae. If we turn to orientation of whole pseudoplasmodia there is a large array of interesting observations. It is, unfortunately, impossible at this stage of our knowledge to know how to interpret the movement of the whole cell masses in terms of movement of the individual amoebae. We shall begin the discussion with orientation to physical agents such as light, heat, and humidity.

Orientation to light in migrating pseudoplasmodia is a striking phenomenon, and Raper (1940a, b, 1941b) was the first to study it in detail. He worked primarily with *D. discoideum*, although phototaxis is also known in many other species. The sensitivity to light is very high, and only minute light intensities are required (Bonner, Clarke, Neely, and Slifkin, 1950). Gamble (1953) showed that he could orient migrating pseudoplasmodia of *D. discoideum* with such light sources as luminescent bacteria and phosphorescent paint. The phenomenon has been thoroughly investigated by Francis (1964), who has even obtained an action spectrum and finds that its peaks resemble those of a flavin or a carotinoid, as is the case for phototropism in higher plants.

Of special interest is the fact that he was able to focus a microbeam of light on different parts of the migrating slug and to show that if he illuminated one side of the tip, the slug would move away from the lighted side (Fig. 18). This is similar to the situation in *Phycomyces* discovered by Buder.[5] The hypothesis is that in normal orientation, the

[5] For a review see G. H. Banbury, pp. 530-579 in *Encyclopedia of Plant Physiology*, Vol. 17/1, W. Ruhland, ed. Springer-Verlag, Berlin, 1959.

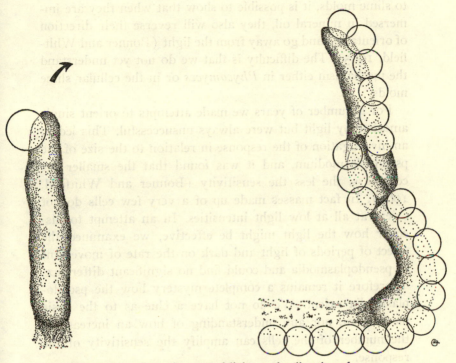

Fig. 18. The effect of a spot of light on the direction of movement of a migrating slug of *D. discoideum*. *Left*: A drawing showing the position which the spot was put on one slug. *Right*: A diagram showing the effect of the spot put successively on one side and then on the other of a slug. Note that the slug tends to turn away from the illuminated side. (Redrawn from D. R. Francis, 1964.)

light is concentrated on the far side by a lens effect of the translucent organism, and that greater growth curvature occurs there. Supporting this hypothesis is the fact that if a sporangiophore of *Phycomyces* is placed in mineral oil, because its refractive index differs from that of air, the light rays are diverged rather than converged by the hypha, and the sporangiophore orients away from light. To come back to slime molds, it is possible to show that when they are immersed in mineral oil, they also will reverse their direction of orientation and go away from the light (Bonner and Whitfield, 1965). The difficulty is that we do not yet understand the mechanism either in *Phycomyces* or in the cellular slime molds.

For a number of years we made attempts to orient single amoebae by light but were always unsuccessful. This led to an investigation of the response in relation to the size of the pseudoplasmodium, and it was found that the smaller the cell mass, the less the sensitivity (Bonner and Whitfield, 1965). In fact masses made up of a very few cells do not orient at all at low light intensities. In an attempt to discover how the light might be effective, we examined the effect of periods of light and dark on the rate of movement of pseudoplasmodia and could find no significant difference. Therefore it remains a complete mystery how the pseudoplasmodia orient. We do not have a clue as to the basic mechanism, nor any understanding of how an increase in the number of the cells can amplify the sensitivity of the response.

Migrating pseudoplasmodia also orient in heat gradients. This was discovered by Raper (1940b), and Bonner, Clarke, Neely, and Slifkin (1950) showed that the phenomenon was one of astonishing sensitivity. In a gradient of $0.05°C/cm$ the migrating pseudoplasmodia of *D. discoideum* migrate readily toward a warmer region. This means that in small pseudoplasmodia the temperature difference between the two sides would be of the order of magnitude of $0.0005°C$.

Potts (1902) was the first to claim that cell masses of *D. mucoroides* were negatively hydrotactic and would orient towards drier regions.[6] Unsuccessful attempts to repeat this

[6] Negative hydrotaxis was suggested, by von Schuckmann (1925) and Harper (1926), to be the cause of aggregation, but, of course, this idea is of historical interest only.

experiment were made by Bonner and Shaw (1957), but recently Shaffer (personal communication) has definitely established that there is a positive hydrotaxis. If dishes are prepared with wells containing substances which control the relative humidity, there is definite orientation towards the wetter regions.

There are reports by Raper (1939, 1940b) that the entire slugs of *D. discoideum* show chemotaxis and will move from an alkaline to an acid region. Despite repeated attempts we have not yet been able to confirm this observation.

All the discussion thus far of various kinds of taxes in pseudoplasmodia has concerned the influence of the larger environment. In recent years there has been considerable interest in the mutual relations between neighboring pseudoplasmodia. This was spurred by a discovery of Rorke and Rosenthal (1959), who were investigating a phenomenon illustrated in a film I made in 1941. If a migrating pseudoplasmodium was cut into segments, the fruiting bodies arising from the segments leaned away from one another. They showed that this was the case even if the pieces were shuffled about so that, for instance, the anterior segment lay between the middle and the posterior segment. This was interpreted as meaning that the orientation was not an intrinsic property of the segment, but the pseudoplasmodia were repelling each other, presumably by some sort of gas.

The matter was pursued in detail by Bonner and Dodd (1962b), and they showed that if two pseudoplasmodia of *P. pallidum* were placed 0.8 mm or less apart, the resulting sorocarps leaned away from each other, the angle being related to the distance. If the two pseudoplasmodia were touching, the sorocarps had an angle of 45° from the vertical. It was also possible to show the response with single pseudoplasmodia by placing them at the junction of two planes of agar: the rising sorocarp would always bisect the angle between the two planes (Fig. 19). The sensitivity of the mechanism could be shown by inserting a glass rod (with a diameter averaging about 0.1 mm) about 0.3 mm from a pseudoplasmodium, which would then lean away a significant degree from the vertical.

A variety of experiments were performed to test the hypothesis that the agent is a gas. The most convincing was the observation that if charcoal was placed near one side of a rising sorocarp, instead of orienting away it would develop directly into it (Fig. 19). Presumably the gas is absorbed by the charcoal, thus reducing the concentration on that side, and since the cell mass tries to escape high concentrations of the gas, it will bend into the absorbent. It is now important to discover the chemical nature of this gas, but thus far we have been frustrated. Shaffer (unpublished) showed that mineral oil would also absorb the gas, and we have considerable evidence that it is not water vapor. But unfortunately this leaves many other possibilities.

Another new and interesting observation is one of Shaffer (1964b), who showed that unaggregated cells of *P. violaceum* or *D. discoideum* have a marked ability to attract the rising sorocarps of *P. violaceum*. This does not appear to be a case of gas absorption, as with charcoal or mineral oil, but appears to involve a gas which is distinct and separate from the one causing repulsion.

One cannot help but be struck, in this discussion of attraction and repulsion of cell masses, by the similarity to the same phenomena in the amoebae. It will be of great interest to learn whether the cells and the groups of cells are attracted by the same substance and whether they are both repelled by another. There is evidence on both levels that repulsion and attraction are mediated by separate mechanisms, but as yet we have not been able to compare cells with pseudoplasmodia.

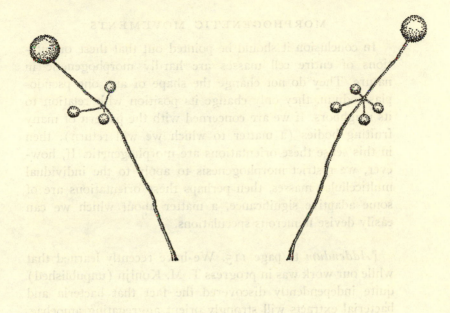

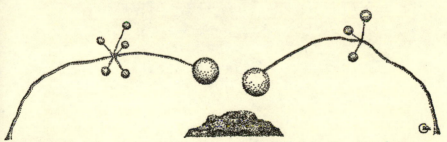

Fig. 19. *Above*: Two fruiting bodies of *P. pallidum* bending away from each other. *Below*: Two similar fruiting bodies bending toward some activated charcoal.

In conclusion it should be pointed out that these orientations of entire cell masses are hardly morphogenetic in nature. They do not change the shape of any one pseudoplasmodium, they only change its position with relation to its neighbors. If we are concerned with the pattern of many fruiting bodies (a matter to which we will return), then in this sense these orientations are morphogenetic. If, however, we restrict morphogenesis to apply to the individual multicellular masses, then perhaps these orientations are of some adaptive significance, a matter about which we can easily devise numerous speculations.

[*Addendum* to page 115. We have recently learned that while our work was in progress T. M. Konijn (unpublished) quite independently discovered the fact that bacteria and bacterial extracts will strongly orient aggregating amoebae. Especially important is the fact that he has made this demonstration using a new quantitative test for orientation that he has devised.]

V. Differentiation

DIFFERENTIATION is generally regarded as the central process in development, for it involves the changes in chemical composition and structure that take place within the cells. It is helpful in discussing the cellular slime molds to make a distinction between two kinds of differentiation.[1] One is that which occurs in a temporal sequence and may not necessarily be connected in any way with a controlled pattern or rigid proportions. This kind of differentiation is especially characteristic of unicellular organisms which undergo a series of changes in time. A good example would be the steps leading to spore formation in a bacterium or a solitary amoeba, and the later steps leading to spore germination. It is this kind of differentiation which has received the special attention of the biochemist in recent years. For convenience we will call this sequential type of differentiation, *differentiation in time* or *temporal differentiation*.

The second kind of differentiation occurs when different parts of an organism take on at one moment in time different characters, different chemical compositions and structures. This kind of differentiation does involve a control of proportions so that the size and distribution of the different regions remain consistent from generation to generation. This kind of differentiation is characteristic of multicellular organisms in which some groups of cells turn into one kind of tissue, while others turn into another. In a very complicated organism these changes may not occur precisely at the same moment, but in many instances the groups do differentiate simultaneously. While this type of differentiation is best illustrated in multicellular forms, it is of course also present

[1] See J. T. Bonner, *Size and Cycle,* Princeton University Press, 1965. I am grateful to Gregory Bulkley for his helpful discussions of these two kinds of differentiation. He has adopted the terms *linear differentiation* and *divergent differentiation* to distinguish between them (Bulkley, 1965).

in unicellular forms which are subdivided into different organelles. For convenience we may call this *simultaneous differentiation* or *differentiation within parts of an organism,* or *spatial differentiation*.

It is important to stress the fact that, as is usual with categories which subdivide developmental processes, the division of differentiation into two kinds is to some extent arbitrary and abstract. Often both kinds of process occur together to such an extent that it is impossible to separate them. Furthermore any simultaneous differentiation involves in one sense a temporal differentiation. One part of a multicellular organism will undergo a series of changes in time which is indistinguishable from what occurs in the temporal differentiation of some unicellular organism. To give an example from the cellular slime molds, a solitary amoeba (e.g. *Hartmanella*) will form a cyst and so will a pre-spore cell in *Dictyostelium*. In *Hartmanella* the whole organism undergoes this change during the course of time. In *Dictyostelium,* some of the cells (the pre-spore cells) undergo this same change, while another group of cells undergoes a different change in time and differentiate into stalk cells. Therefore simultaneous or spatial differentiation involves two or more parcels of temporal differentiation that occur at roughly the same time. Simultaneous differentiation is thus a more complex differentiation; it is at a higher level and involves the grouping of more than one individual temporal differentiation. At this advanced level of complexity our interest is not solely in the sequence of steps leading to the changes involved in individual temporal differentiations, but also in the control mechanisms which make the individual temporal differentiations harmonious and proportionate with one another. In other words, we are concerned at this higher level with the mechanisms coordinating the separate temporal differentiations.

We plan to discuss first the differentiation in time. We have already, in Chapter II, presented some of the descriptive background, and now we shall examine what we know of some of the sequences of steps which comprise temporal differentiation. There are three kinds of problem, the first and last of which will concern us. A sequence of steps may lead to the construction of some key substance, such as cellulose—key in the sense that without it the development would utterly fail. A sequence of steps may, on the other hand, lead to some substance which is not critical—the absence of which in fact does not alter development. An example of the latter would be the production of the purple spore pigment (presumably melanin) in *D. purpureum*, which can be inhibited by phenol, as was demonstrated by Whittingham and Raper (1956), without any effect on the final fruiting body except the trivial one of loss of pigment. This type of sequence will not be examined here.

In these first two kinds of problem one is concerned primarily with an internal sequence of events, and therefore the stimulus upon which any one step depends is simply whether or not the previous step in the sequence has taken place. In the third and last kind of problem we are concerned with those steps in a sequence that require an external stimulus, or that are at least in some way affected by the environment. In lowly organisms the cues are very apt to be environmental, for it is essential for survival that the developmental processes be geared to changes in the outside world. In the evolutionary trend toward large and more complex forms, there is increase in internal stability independent of the environment (e.g. homeothermy), and a greater number of the trigger mechanisms for the various stages of differentiation are internal (e.g. induction).

In the discussion of simultaneous differentiation, one of the first important considerations is the size of the cell mass.

A whole field of amoebae is carved up into a number of aggregates, and we must examine what is known of the mechanisms controlling the size of the groups.

Once a cell mass has formed, then the proportion of spore to stalk within it becomes of prime concern. There is not only the important question of how such a mechanism operates, but also how it can adjust to sudden changes in size if the cell mass is bisected. It will not be possible to answer these grand questions, but what facts are known and what interpretation we can put on them will be discussed.

1. Temporal differentiation: the biochemical approach

One of the staggering thoughts in developmental biology is the number of biochemical reactions or steps that must take place to bring about even the simplest and briefest developmental change. This means that the biochemist cannot hope ever to learn or catalogue all of them, and he therefore seeks those chemical steps which are of key significance in a particular developmental phase. A good approach, often used, is to find first a key stage and then assault this with our modern biochemical tools. Another method frequently used is to look for the presence or absence of some particular substance which is known from some other organism to be of metabolic significance. Sometimes the road is so new and the possibilities so limited that a certain technique of known capabilities and ease of manipulation is applied to a variety of stages of development.

All these approaches yield information, sometimes large amounts of it, and we are then faced (as we are at this moment) with the problem of putting it together into some meaningful story. Unfortunately if we do this right now we cannot escape the conclusion that we do not yet know enough for any comprehensive understanding. We do have some isolated bits of information of great interest which we shall briefly review here, but we shall do so in the broadest possible terms, trying to isolate a few new insights that have been gained from a biochemical approach. For the details the reader is urged to consult the reviews of Gregg (1964, 1966a) and Wright (1964).

By way of introduction a few points should be made concerning the metabolism of the cellular slime molds, even though this is a problem that may be fairly far removed from the biochemistry of development. The general consensus is that the amoebae are at all times aerobic.[2] Wright and Anderson (1958) have described certain enzymes concerned with metabolism and have shown that certain dehydrogenases increase as development begins. Takeuchi (1960) showed a similar increase for succinic dehydrogenase, and a corresponding decrease for cytochrome oxidase. In this latter case the changes were associated with the pre-spore area. Furthermore they are correlated with a change in the structure of the mitochondria, which also occurs at the interphase period between feeding and aggregation. The great difficulty is the question of whether these changes are the result of the significant biochemical changes or whether they themselves are responsible for the changes, as Wright (1960) suggests.

The cellular slime molds are in some ways similar to embryos in general in that they store up their energy reserves before they begin their morphogenesis. Here there is no great accumulation of yolk, but merely well-fed amoebae, some still containing food in their food vacuoles. In the growth phases, and during aggregation and the beginning of migration, Gregg, Hackney, and Krivanek (1954) showed that the energy source did not involve the breakdown of proteins, but once the immediate reserves were depleted, there was a great destruction of protein, presumably as an

[2] There is not complete agreement on this point; for a discussion see Gregg (1964, p. 648).

energy source for the later development. This is interesting in view of the fact that the later stages of development involve a considerable synthesis of polysaccharides in the form of slime and especially cellulose for the stalk and the walls of the spores. There are a number of studies on antimetabolites, amino acid analogues, and increases in enzymes involved in carbohydrate synthesis which support this picture (review: Gregg, 1964, 1966b).

We can therefore begin to think of some of the morphological changes in biochemical terms. Let me give a few specific examples. White and Sussman (1963) showed that an acid mucopolysaccharide appears during aggregation and reaches a level, in dry weight, of 1 to 2 per cent of the total by the end of culmination. It is apparently solely associated with spores and not with the stalk at all. They further demonstrated that the enzyme responsible for the synthesis of this substance is a uridine di-phosphate galactose polysaccharide-transferase which appears about an hour before the polysaccharide itself can be demonstrated. It rises rapidly in amount as development proceeds, then is lost by excretion to the outside with the cessation of synthesis (Sussman and Osborn, 1964; Sussman and Lovgren, 1965). More recently Sussman and Sussman (1965) have been able to inhibit the synthesis of the enzyme specifically with actinomycin D, which is known to block RNA synthesis. Furthermore the actinomycin is successful in blocking the enzyme production (and hence the synthesis of the polysaccharide) only if it is applied four to five hours before the usual time of first appearance. It would seem, from this intriguing story, that we have some inkling of the sequence of steps for this particular enzyme, (as is true for some other developing systems), and it is hoped that soon others will be elucidated in the same fashion. Clegg and Filosa (1961), Ceccarini and Filosa (1965) and Ceccarini (1966) have the beginnings of such an analysis in the trehalase-trehalose synthesis which was discussed earlier in our examination of spore germination.

It is impossible to tell at the moment if a series of such isolated sequences of steps will give us a clearer insight into development or merely a complicated catalogue of facts. It would seem to me that there is every reason to be optimistic. Admittedly what we have is in a sense pure description, but there is no better material for finding regularities and significant generalizations than a series of penetrating empirical observations. In fact it is hard to conceive of any other approach at the moment than the descriptive biochemical one. This approach is now made within the conceptual framework of DNA-RNA-protein synthesis, and for the time being this has been helpful. If it is ever to change, or if it is ever to give us a complete picture of development, it will only be after the grubbing for many more biochemical facts.

There are, of course, other tools than the ones used in the studies just mentioned. For instance, good use has been made of immunological techniques; by this method it is possible, in a very sensitive fashion, to detect the appearance of new macromolecules and the disappearance of old ones. This is well illustrated in Gregg's (1961) study using immunoelectrophoretic techniques in which he showed that during the course of development of *D. discoideum* there is a continual appearance of new antigens and a disappearance of some that were present from the beginning. Also Sonneborn, Sussman, and Levine (1964) were able to demonstrate a new antigen appearing apparently at the surface of the cells, just at the time of aggregation. Takeuchi (1963) and more recently Gregg (1965) have made effective use of fluorescent antibody methods to show the appearance of new antigens, and their work will be considered presently. It is, of course, imperative to find out what these substances are chemically, and to understand their synthesis, as was done for the mucopolysaccharides and trehalose.

But of course, the most important goal of all will be the understanding of what turns one substance off, and the other on. We imagine the control mechanism as the golden key, and it is hoped that with new knowledge, new techniques, and new ideas, we will ultimately achieve this goal for the cellular slime molds. Often, as one peers intently at the details of a biochemical solution of development, one loses track of precisely what it is that is controlled, what it is that needs explanation. Admittedly all our problems cannot be solved at once, but it is essential to keep in mind at all times the nature of the problem. It is more than solving numerous DNA-RNA-enzyme-product sequences; more even than knowing what controls any particular sequence. The remainder of this chapter will be devoted to this necessary perspective. We will begin with a simple discussion of the external factors which guide and stimulate different stages of development in temporal differentiation, and from there we will lead to the all-important problems of the regulation of size and proportion during spatial differentiation in the cell community.

2. Temporal differentiation: environmental triggers

Since the cells of slime molds are at all times at the direct mercy of the elements, it is not surprising to find that almost all the stages of development are to some extent environment-dependent. The most obviously so is spore germination, discussed previously. The conditions which provoke germination (as with other organisms) are the conditions which are favorable for growth. The key factors are temperature and humidity, and these can be equally limiting for both germination and growth. Also it will be remembered that spores produce a germination inhibitor in the environment which suppresses germination in dense spore populations. This, of course, does not affect the development of the organism, since a new generation can be produced by a few or even only one amoeba.

That starvation is the prime stimulus for the aggregation stage has been known for a long time and is discussed by Potts (1902), Oehler (1922), von Schuckmann (1925), and Raper (1940a). It is even possible to prolong the growth phase by constantly adding bacteria, as has been shown by Potts (1902) and Raper (1940a). The starvation stimulus for fruiting is not confined to the cellular slime molds, but is a widespread phenomenon. It is found in the myxomycetes and in the majority of the filamentous fungi. It would seem to make rational sense, for if the food supply is giving out, there is no safer turn than encapsulating the cells in resistant bodies to tide the organism over adverse periods.

The onset of aggregation is also to a minor extent affected by humidity, heat, and light. This was first discovered by Raper (1940a), who showed that aggregation occurs two to four hours sooner if the humidity is decreased, the temperature raised, and the light turned on. It is quite possible that the temperature and humidity effects are one and the same, but the light effect is probably separate. It should be added that Raper also found that these environmental stimuli produced more numerous and smaller aggregates, but we shall consider this in the discussion of size further on.

We do not know how these agents act, although Kahn (1964b) has some evidence that light blocks a substance inhibiting center formation (to be discussed presently). Shaffer (1961b) demonstrated that light produces a burst of new founder cells, and it is possible that the mechanism by which this operates may also be the blocking of a center inhibiting substance. Kahn also showed that in *P. pallidum* the stimulus can be produced by a light period of very short duration early in the phase before aggregation. This fits in with the observation of Shaffer (1958) that aggregation in *D. discoideum* can be to some extent synchronized by alternating periods of light and dark.

There are no experiments delving into the problem of how humidity can affect the onset of aggregation, but there is one curious recent observation on temperature by Konijn (1966). He showed that if two groups of cells are separated from each other on an agar surface, the distance over which one group can attract the other in aggregation is much greater at low temperatures than at high temperatures (50 per cent attraction will occur over a distance 1535 μ at 13°C and 955 μ at 24°C). It is not known if this effect is on the stability of acrasin or on the sensitivity of the response. It is interesting that factors which produce an increase in the attraction distance should lengthen the time of onset of aggregation.

If we now turn to the factors which control the duration of migration, it must be understood that we are dealing here in effect with the factors which trigger the beginning of culmination, the beginning of the final differentiation of the spores and the last of the stalk cells. The first clue to the problem came from a series of experiments in which we demonstrated that the concentration of solutes in the agar, all other factors being equal, had a profound effect on migration in *D. discoideum* (Slifkin and Bonner, 1952). The higher the concentration of solutes, the shorter the length of migration.[3] On plain, non-nutrient, two per cent agar the migration averaged ten days, with one extreme case of migration for twenty days. In some instances, if the slug was small from the beginning, it would just migrate until it disappeared, presumably because of the loss of cells in the slime track and

---

[3] Electrolytes were found to be slightly more effective than non-electrolytes at equivalent molar concentrations. There still is no satisfactory explanation of this difference.

the consuming of its energy reserves in these optimal conditions for migration.

The same technique has more recently been used with *D.
mucoroides* with a similar result, but because *D. mucoroides*
produces a continuous stalk during migration, it has certain
advantages for such a study (Bonner and Shaw, 1957). In
fact, it was possible to show that very small changes in humidity had a profound effect on migration; the slightest decrease would cause migration to stop and spore differentiation
to commence. If a migrating mass is lifted off the surface
of the agar there is apparently a sufficient difference between the relative humidity near the surface of the agar and
that a few millimeters away to stimulate the beginning of
the final differentiation. This was first discovered quite by
accident in an experiment in which a number of migrating
pseudoplasmodia of *D. mucoroides* were on a non-nutrient
agar, moving toward a small light source. Through some
error, one of the culture dishes was placed upside down, and
as the stalk produced by the cell masses lengthened, the weight
of the masses became excessive, causing them to swing down
as though they were hanging from an agar ceiling. Almost
as soon as they had become detached from the agar surface
they ceased migration. This effect could be repeated in right-
side-up culture dishes by raising the light source at an angle
above the petri dish. In order to migrate toward the light
the cell masses now had to rise upward at an angle of about
30°. Of course they kept falling back onto the agar when their
stalks became too long, but on the average, culture dishes
that were at such an angle from the light source fruited much
sooner than plates which were on the same plane with the
light. The point was checked in other ways, especially using
solutions of sulphuric acid which lowered the relative humidity of the atmosphere over them, and all the results confirmed the view that relative humidity was of extreme importance.

This view conforms with the opinion of Raper (1940b), who showed that by lifting the cover off of the culture dishes for short periods it was possible to induce migrating pseudoplasmodia of *D. discoideum* to enter into the culmination phase. Also, it should be mentioned that Potts (1902) was keenly aware of the importance of relative humidity, although he considered it to play a role in transpiration. He believed that in *D. mucoroides* stalk formation was promoted by a slight decrease in relative humidity so that transpiration could operate, and that saturated conditions inhibited fruiting. We now know that his experiments under saturated conditions are more likely interpreted in terms of the accumulation of toxic or inhibitor substances (Bonner and Hoffman, 1963). Potts himself showed that if amoebae were grown in a sealed container with air bubbled through a water trap over the culture, good development resulted, but he was so convinced of his transpiration hypothesis that he sought other explanations for this experiment.

Our results with relative humidity on *D. mucoroides* led us to suggest that the reason that increasing the solute concentration reduced the extent of migration was because it effectively lowered the humidity at the surface of the agar. Since the organisms give every evidence of being extremely sensitive to desiccation, there is no reason why the drying cannot be effected osmotically. Also, another set of experiments of Raper (1940b) might have a similar explanation. He showed that an increase in temperature resulted in the cessation of migration. That is, if, during the migration phase, a culture dish was taken from a low temperature and brought to a higher one, migration would stop. This temperature rise would, of course, lower the relative humidity, which would be effective in promoting culmination. If the reverse was done, that is, the temperature lowered, Raper found prolonged migration. At constant high temperature he reports short periods of migration, but we were able to show that if sufficient precautions are taken to prevent evaporation, then migration may occur for long periods at temperatures as high as 30°C for *D. discoideum* and 32°C for *D. mucoroides* (Bonner and Shaw, 1957).

Whittingham and Raper (1957) have shown that the humidity requirements of *D. polycephalum* are rather special, although in general they follow the same principle. At first great difficulty was encountered in inducing the migrating pseudoplasmodia of *D. polycephalum* to culminate at all; they continued to migrate indefinitely. By accident it was found that in the presence of the mold *Dematium nigrum,* fruiting occurred readily, and by a careful analysis of this phenomenon, using solutions of different relative humidities, Whittingham and Raper were able to show that the effect of the *Dematium* could be imitated by lowering the humidity. Here again, high humidity favors migration, but culmination requires even lower relative humidities than in any of the other known members of the Acrasiales.

In an experiment with *D. mucoroides* under optimum high-humidity conditions we could keep the migration going for such long periods of time that it was possible to produce greatly extended fruiting bodies—far longer than any previously reported. Our record was a length of 22 cm, and I feel sure we could better this if we had a larger culture dish. Great lengths were obtained not only with some strains of *D. mucoroides,* but also with certain strains of *D. purpureum. P. violaceum*, in one strain, produced lengths of 8 cm, and even some isolates of *P. pallidum* could be considerably elongated in this way. In *D. mucoroides* the phenomenon is especially striking since the original cell mass is not especially large, and so with continued migration the majority of the cells enter the stalk, leaving a minute sorus at the end of an absurdly long stalk.

In some unpublished work, Shaffer has shown that it is possible to make *P. violaceum* differentiate almost totally into stalk cells. He places mineral oil over agar plates containing the developing slime mold, and if this is done correctly the developing stalks become trapped in a small, flat droplet of water which lies below the oil. In this condition the pseudoplasmodia migrate around and around many times in their prison, forming great masses of curled stalk, with often little or no spores. Shaffer interprets this as being another example of high humidity favoring prolonged migration, for in this case the cells are forced by the oil-water interface to stay in contact with the water drop.

In concluding this section on the external triggers, some discussion is necessary of the recent work on the external conditions which lead to macrocyst formation in *D. mucoroides*. It has been shown independently by Hirschy and Raper (1964) and Weinkauff and Filosa (1965) that macrocysts form only in the dark; they are inhibited in the light. It was also shown by Weinkauff and Filosa (1965) that macrocysts would form in greater abundance in airtight, sealed tubes than in ones with cotton plugs. Recently Filosa (personal communication) has shown that this is probably due to an increase in humidity rather than to an accumulation of a volatile metabolite as originally suspected. In fact if the cultures are slightly submerged in water, he can make all the aggregations turn into clusters of macrocysts rather than fruiting bodies, and this will occur even in the light. The water effect will override the light effect.

3. Spatial differentiation

The simplest and most direct kind of simultaneous or spatial differentiation that can be observed in the cellular slime molds is the organization of a whole field of amoebae into aggregates. As will be clear, there are factors which control both the number of aggregates and the size of the aggregates, and number and size have some interesting relations one with another.

This story begins with a certain amount of controversy. It was the original contention of Sussman (1958 et seq.) and his co-workers (Ennis and Sussman, 1958b) that there were particular cells that differentiated into "initiator cells" and that these were in a fixed proportion to the number of other cells in a population. For instance, in *D. discoideum* one cell in every 2,100 was supposed to be such a cell; for other species and strains the ratio was different. The evidence against such a fixed differentiation by numbers is now overwhelming, and the discussion of it lengthy (Gerisch, 1961b; Konijn and Raper, 1961; Bonner and Dodd, 1962a; Shaffer, 1962). In order to avoid a recapitulation of this polemic let me cite one experiment which refutes the notion and is in itself of considerable interest.

Many workers had made attempts (including elaborate experiments in our laboratory) to confine a small number of pre-aggregation cells in a restricted area. These experiments always failed: the cells wandered off over any kind of barrier that could be devised. A solution to the problem was finally achieved by R. R. Sussman (cited in Ennis and Sussman, 1958b): if a drop of water containing cells was placed on thoroughly washed agar, the cells did not wander beyond the original confines of the drop. This technique was further investigated by Konijn and Raper (1961), who found that the amoebae could be dispensed in drops of standard salt solution; this increases their viability, yet they remain confined in the zone of the original drop. Konijn and Raper concluded that the cells are probably prevented from leaving the drop area because the washed agar outside is hydrophobic in nature.

Using this improved method, Konijn and Raper systematically tested small populations of cells, and they discovered that the success of aggregation in these small drops far exceeded what could be expected on the basis that one cell in every 2,100 was capable of initiating aggregation (Plate 8). In fact if the number of cells per drop was of the order of 200, then aggregation occurred in 100 per cent of the drops. If the number of cells per drop was of the order of 100, aggregation took place in approximately 75 per cent of the cases (within the range of 50 to 90 per cent). The important point is that aggregation is not dependent upon there being a fixed ratio of initiator cells to other cells in a population. What then are the factors which determine the number of aggregates in an area and the size of the aggregates?

The first to make quantitative measurements were Sussman and Noël (1952; see the more recent review of Sussman and Sussman, 1961). They dispensed large drops of amoebae on agar and measured the total number of cells in a drop, the density of cells per unit area once the cells were dispensed, and the final number of aggregations within the entire area. If cell density is plotted against the total number of aggregates, an optimal cell density is evident; above and below this density there are fewer fruiting bodies. Sussman and co-workers consider this a useful method of testing what they refer to as "aggregation performance"; for instance Bradley, Sussman, and Ennis (1956) showed that histidine permitted aggregation to occur at lower densities than normal, while adenine and guanine inhibited aggregation so that it occurred only at higher cell densities.[4] Crude cell extracts greatly increased the number of centers formed at the optimum cell density.

It is most unfortunate that in all these detailed quantitative studies the measurement of the total area of the drop was not

[4] These studies have been confirmed and pursued further by Krichevsky and Wright (1963) and Krichevsky and Love (1964a, b).

also included. It has been shown by Bonner and Dodd (1962a) that a parameter of key importance is the density of the fruiting bodies or, to put it in inverse terms, the size of the aggregation territory. Were the drop size known this could be easily calculated for all those tests involving "aggregation performance."

The notion of territory size is important in its own right, and not simply because it is a convenient measure. It was shown for a number of different species that the aggregation territory remained constant over a wide range of cell densities (Fig. 20). We were particularly fortunate in these early experiments (Bonner and Dodd, 1962a) in using a technique that gave such constant results, for subsequently we found that, while the principle is sound, different techniques on particular species will produce greater variability (Bonner and Hoffman, 1963). The important point, however, is that there is clearly a method by which the determination of the centers is achieved, and within a great range this is independent of amoeba density.

The distribution of the centers themselves is non-random; obviously one possible explanation is that there is some way in which the centers are controlling one another. The evidence that there is a center inhibitor is now clear cut and comes from two independent studies. Long ago Arndt (1937) described the fact that centers would often disintegrate, but no analysis of this possible center suppression was made for many years. Shaffer (1961b) discovered the most interesting fact that in *P. violaceum* certain cells (founder cells) in a population rounded off and individually became the focal points, the keystones for aggregates (Plate 2).[5] The number of these cells is not fixed in proportion to the total number of cells, and if one is eliminated, others will

[5] Founder cells have now been described in *P. pallidum* by Francis (1965) and in *D. minutum* by Gerisch (1964b). It is still uncertain whether or not they are present in other species (Shaffer, 1962).

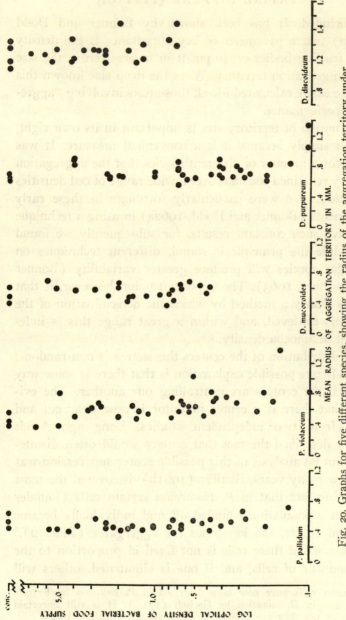

Fig. 20. Graphs for five different species, showing the radius of the aggregation territory under conditions of different bacterial food densities. The top points labeled *conc.* are on 2% agar without streptomycin, and a heavy layer of bacterial paste is smeared evenly over the surface. Each point is an average of 5 squares (0.102 cm²) on one culture dish. (From Bonner and Dodd, 1962a.)

arise. The significant point here is that if the founder cell is not removed, none of the neighboring cells turn into founders. Later, Shaffer (1963a) studied this inhibition specifically, and by opposing two layers of agar, one of which contained aggregates and was therefore at a more advanced stage than the other. The older aggregates were successful in suppressing the formation of new centers in an area surrounding them.

The other demonstration of inhibition was first made with *D. mucoroides*; the pre-aggregation amoebae were placed in agar in a closed container, with the result that aggregation was totally inhibited (Bonner and Hoffman, 1963). However, there was no inhibition in the stoppered vessel if an absorbant such as charcoal or mineral oil was added. The indications are, therefore, that for this species the inhibitor is a gas, but apparently this is not true for *D. discoideum,* where the inhibitor acts only in solution. Again these are matters which can be satisfactorily solved only when the chemicals are isolated and known.[6] The evidence for the existence of a center suppression substance is, however, good and comes from a number of subsidiary experiments. If, for instance, two agar surfaces containing aggregating cells are opposed (with an air space between them), aggregates do not form opposite one another but are appropriately spaced. For this reason we called the substance a "spacing substance," but the "center suppressor" of Kahn (1964b) or "center inhibitor" are equally clear and satisfactory terms.

It should be added that Kahn (1966) has recently shown that some of the examples of non-random spacing can be fully explained on the basis of the timing of the appearance of centers. If all the centers develop rapidly, as they tend to do in dense populations, they will approach a random distribution (except in very dense populations where they

[6] It was first thought (Bonner and Hoffman, 1963) that the gas was CO_2, but in view of recent work this seems most unlikely.

cluster). He offers two possible hypotheses for the fact that sparse populations are non-random in their center distribution. The simplest is that if one center appears much sooner than its neighbors it will subtract many amoebae from the surroundings, so that centers that appear later are necessarily a certain distance away. The other explanation is that there is an inhibitor which takes time to diffuse out. Clearly the former phenomenon must always be operating, and in many cases there is evidence of a spacing substance as well.

There are a number of interesting properties of the spacing substance. In the first place not all species have the same sensitivity to it, and it has already been mentioned that some species are gas-sensitive and others not. In *D. purpureum* the spacing mechanism is particularly effective regardless of amoeba density, yet center formation, once it has taken place, cannot be reversed by harvesting the cells, mixing them, and replating them. The original centers stay fixed all through this rather hard treatment. On the other hand, in *P. pallidum* the suppression of centers already formed is easily achieved and can be readily done after a second mixing of the cells. Another interesting discovery is that of Kahn (1964b), who showed that the reason the presence of light results in a reduction of the territory size is that light in some way blocks the action of the center inhibitor or suppressor.

It is evident from the foregoing that there must be at least three separate processes connected with aggregation. One is the process of center formation, which in some species is characterized by an obvious founder cell. The second is the center inhibition mechanism, which is responsible for controlling the spacing of the aggregates. These two processes, working together, manage to bring order out of randomness. The third essential process is amoeba orientation. Gradients of acrasin are somehow formed, which have their high points at the centers; center formation and acrasin emission must be related. The cells at this stage also enter their phase of high sensitivity to acrasin, so that the gradients orient the individual cells and cause the formation of polar streams that lead to the centers. As always in discussions of this kind, we revert to the question of whether these three processes are utterly distinct biochemically or whether there are some substances that can accomplish more than one of the effects, but we cannot satisfactorily answer this question until the substances have been isolated.

There is another phenomenon, described by Gerisch (1964b), that also requires explanation. He noted that in thick cultures of *D. minutum* a center without streams formed first, but that subsequently this center disintegrated and the cells wandered out in streams and then formed secondary centers about the periphery (Fig. 21). At even higher cell densities, the central center may re-form after partial disintegration and produce fruiting bodies along with the secondary centers. Gerisch considers the formation of secondary centers an induction, but the only evidence at the moment is their description. It is conceivable that a balance between center forming tendencies, center inhibiting tendencies, acrasin gradients and their timing, as well as the timing of the sensitivity to acrasin, could alone be sufficient to account for the pattern. Again we need actually to do our experiments with the specific substances and find exactly what is the contribution of each.

In the discussion thus far we have concentrated on the pattern of aggregation taking place over a wide area, and now we shall examine the related problem of the size of the pseudoplasmodium. It is related because clearly if territory size remains constant over a wide range of amoeba densities, then the size of the pseudoplasmodium will vary directly with amoeba density. To put the matter in its simplest terms, if the aggregation territory is 3 mm² and if

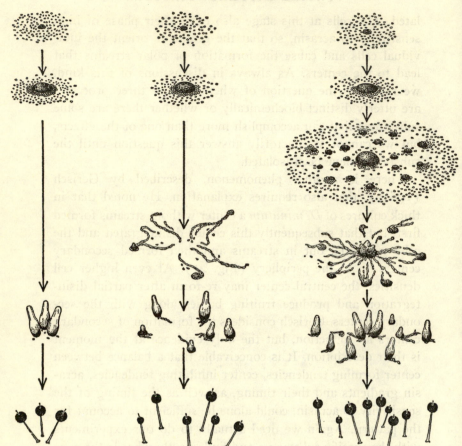

Fig. 21. Drawing showing the various types of aggregation pattern in *D. minutum*. With increasing concentration of amoebae (from left to right) peripheral secondary centers may form, and in some cases (middle) the primary center may disintegrate. (From G. Gerisch, 1964b.)

there are 10 amoebae in that area, the resulting aggregate will produce a fruiting body of 10 cells. If, on the other hand, there are 1,000 cells every 3 mm², then the fruiting bodies will consist of 1,000 cells. In such a study Bonner and Dodd (1962a) found it possible to produce exceedingly small fruiting bodies by reducing the amoeba density, the smallest being one of *P. pallidum* of 7 cells (4 spores and 3 stalk cells. Fig. 22).[7]

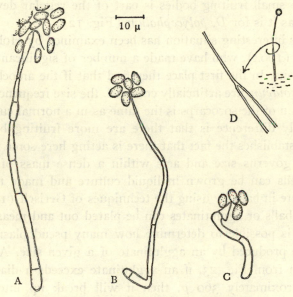

Fig. 22. Camera lucida drawings of small sorocarps. A. *D. purpureum* with 7 stalk cells and 45 spores (not shown). Note the wisp at the end of the stalk. B. *P. pallidum* with 5 stalk cells and 6 spores. C. The smallest sorocarp obtained. It is *P. pallidum* with 3 stalk cells and 4 spores. D. *D. lacteum* showing a small stalk which midway becomes acellular. (From Bonner and Dodd, 1962a.)

Size of the cell mass is determined in this way only at low cell densities. Furthermore the size is in essence a trivial by-product of the spacing control, and the spacing does not possess within itself a pseudoplasmodium size control mechanism. There is another kind of size control that is internal. This has been suspected ever since it was noted that in small

[7] Sussman (1955a) was also able to obtain small fruiting bodies with his Fruity mutant of *D. discoideum*.

species at high cell densities a center will break up into a cluster of fruiting bodies (e.g. *D. minutum*, Fig. 21; *Acytostelium*, Fig. 5). In fact in *Polysphondylium* such a breaking-up into small fruiting bodies is part of the regular development, as it is for *D. polycephalum* (Fig. 12).

This interesting situation has been examined by Hohl and Raper (1964), who have made a number of significant contributions. In the first place they find that if the amoebae of *D. discoideum* are artificially crowded, the size frequency distribution of the sorocarps is the same as in a normal culture; the only difference is that there are more fruiting bodies. This establishes the fact that there is acting here some factor which governs size and acts within a dense mass of cells. The cells can be grown in liquid culture and made to agglutinate in balls by using the techniques of Gerisch (1960). These balls or agglutinates can be plated out and measured, and it is possible to determine how many pseudoplasmodia will be produced by an agglutinate of a given size. As can be seen from Fig. 23, if an agglutinate exceeds a diameter of approximately 360 μ, then it will break up into two sorocarps; this size limit is referred to by Hohl and Raper as the "critical mass."

These workers also tested some mutants which produced small fruiting bodies in culture. Some of these mutant strains remained small in crowded as well as in sparse conditions, while others became indistinguishable from the wild type when crowded. One may assume that in the former the critical mass is genetically distinct, while in the latter there is some factor which is diluted in sparse cultures. The existence of such a factor is also indicated by Hohl and Raper's observation that there is a mutant which fails to aggregate—an "aggregateless" mutant—unless the amoebae are concentrated, when they not only aggregate but form sorocarps of normal size.

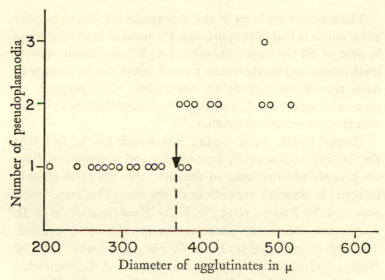

Fig. 23. Demonstration of the critical mass. The diameter of agglutinates of *D. discoideum* plotted against the number of pseudoplasmodia that develop out of such masses. Note that if the mass exceeds a certain size (arrow) the number of pseudoplasmodia doubles. (From Hohl and Raper, 1964.)

In their discussion of this work they point out that there are many parallels in experimental studies on animals where there is also a critical mass or size limit to the units of organization.[8] Size control in the cellular slime molds therefore belongs to the general class of control problems and is of considerable interest. In this case it is not so easy to suggest how we can uncover the mechanism. More is necessary than attempting to isolate a substance; some feedback mechanism of a fairly sophisticated nature must be involved. All we know is that size can be genetically determined. Perhaps if we understood what the deficiency is in those small mutants that with crowding become large, we would make a step toward understanding the problem. Size regulation is undoubtedly a question of great interest and importance.

[8] For references see their paper (Hohl and Raper, 1964).

The supreme problem in the differentiation of the cellular slime molds is that of proportions. Proportion in the cell mass is first of all the best example of a differentiation control mechanism, and furthermore the cell mass can be shown to have remarkable powers of adjustment after surgery, or powers of regulation. Let us begin by examining the evidence that proportions are controlled.

Harper (1926, 1929, 1932) was struck by the fact that the fruiting bodies of *D. mucoroides* and *Polysphondylium* are roughly proportionate in the sense that the ratio of stalk to sorus is constant regardless of the size. The same point was made by Raper (1935) in his original description of *D. discoideum*, where the phenomenon is even more marked. The first quantitative study on *D. discoideum* was made by Bonner and Eldredge (1945), but they used the somewhat unsatisfactory method of measuring stalk lengths and sorus diameters.[9] Later, Bonner and Slifkin (1949) made a detailed study on the same species in which they estimated the volumes of the stalk and sorus and stalk volume was plotted against sorus volume (Fig. 24a). The sorocarps were prepared in whole mounts and measured in a precise fashion. The difficulty with the method was that it assumed that in all the preparations, and in all the sorocarps, the volume of the cells in both stalk and spore was constant. Nevertheless the results did show an exceedingly rough proportionate relation between stalk volume and sorus volume. In order to circumvent the cell volume problem the dry weight of sori and stalks was determined, using a sensitive quartz helix balance according to the technique of Gregg and Bronsweig (1956a). In this case 30 individuals had to be measured at one time in order to obtain a significant reading, and the result again showed proportional development, but the scatter of the points was if anything greater than in the previous study

[9] Their results have recently been confirmed by Hohl and Raper (1964).

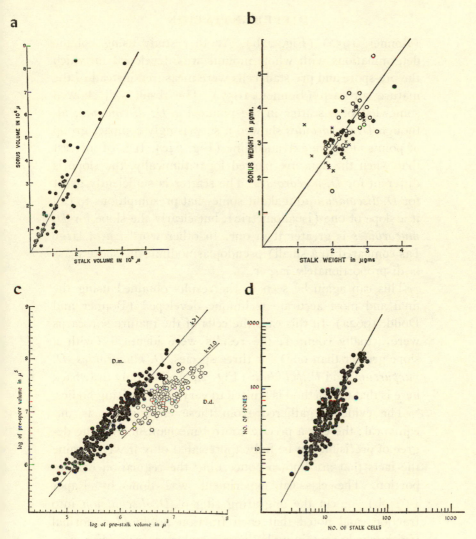

Fig. 24. Graphs showing four different methods of measuring proportions in the cellular slime mold. a) Volume measurements of mature sorocarps of *D. discoideum.* b) Dry weight measurements of same (each point is an average of 30 fruiting bodies. c) Volume measurements of pre-spore and pre-stalk cells of *D. mucoroides* and *D. purpureum.* d) Cell numbers in small mature sorocarps. These are a composite of *D. purpureum, D. mucoroides,* and *P. pallidum* as they showed no discernible difference among them. (See text for references.)

(Bonner, 1952) (Fig. 24b). Another study using volume determinations with whole mounts was developed in which the pre-spore and pre-stalk cells were measured instead of the mature sorocarp (Bonner, 1957). The result still showed somewhat of a scatter in the points for *D. discoideum*, although *D. mucoroides* showed a surprisingly compact group of points along an extended line (Fig. 24c). It is of interest that when the data are plotted logarithmically, the slope is different for *D. mucoroides*. The scatter is sufficiently great for *D. discoideum* to make it somewhat presumptuous to call it a slope of one (isoallometric), but clearly the slope for *D. mucoroides* is greater than one. In other words, in a large (as compared to a small) pseudoplasmodium the spore mass is disproportionately larger.

This can again be seen in the results obtained using the final and most accurate technique developed (Bonner and Dodd, 1962a). In this case the cells of the mature sorocarps were actually counted; the results were identical (with a slope greater than one) for three species, *D. mucoroides, D. purpureum,* and *P. pallidum* (Fig. 24d). The only drawback here is that the method is limited to very small fruiting bodies.

The evidence gathered from these experiments is unequivocal: there is a precise control mechanism. But the degree of precision can be fully appreciated only if we examine the facts that are known concerning the regulation of proportion. The classical experiment was done by Raper (1940b); he cut the migrating slug of *D. discoideum* into fractions and noted that each fraction produced a normal fruiting body containing both stalk and spore cells (Fig. 25). The anterior fraction was special, in that it took a period of migration before the proportions became normal; that is, regulation in the tip takes more time, and if the culmination occurs right after the operation the sorocarp has a disproportionately thick stalk and a small sorus. Given enough

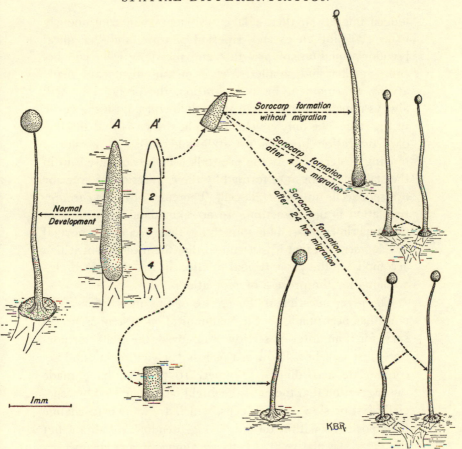

Fig. 25. Comparison of the fruiting of entire cell masses of *Dictyostelium discoideum* with different fractions of the same. If apical fractions fruit immediately they show abnormal proportions, but with some migration the normal proportions are resumed. (From Raper, 1940b.)

time, any cell which was previously destined to become a stalk cell may be made to turn into a spore, and vice versa.

In the case of *D. mucoroides* it is obvious that some such process is continuously operating. In observations on the pre-stalk and pre-spore cells (Fig. 24c), it is found that if the conditions are sufficiently humid and favorable for prolonged migration, then a large cell mass will continuously make stalk for an extended period of time, and the apical pseudoplasmodium of pre-stalk and pre-spore cells will become smaller and smaller. Yet if measurements are made at any one moment in its migration, the proportions will always remain constant, and the points will fall on the curve in Fig. 24c. The only conclusion to be drawn is that the line of demarcation between pre-stalk and pre-spore is slowly moving posteriorly. As the pre-stalk cells become used up in the stalk, new ones are formed by the conversion of anterior pre-spore cells into pre-stalk cells. Therefore in *D. mucoroides* regulation is not something to accommodate the whims of the experimentalist, but something that occurs in normal development. In fact it is assumed that the mechanism which accomplishes the proportional relation in the first place also accounts for the process of regulation.

The great question is the nature of this mechanism. The point has been made that a communication system is necessary, for if an anterior section is removed, the posterior portion must somehow know of this loss (or vice versa) and accommodate accordingly (Bonner, 1957). We have made the suggestion that the message might be transmitted by especially fast or slow moving cells, which are known to change their positions within the migrating pseudoplasmodium, but this is too speculative. The cells are aligned in a polar fashion, and there are many other ways we could postulate by which such messages might be transmitted forward or backward in the pseudoplasmodium. The only thing of which we can be certain is that there must be some sort of message to achieve the remarkable coordination and control that we observe.

There has been a beginning of an attempt at a biochemical analysis of the problem of regulation. By repeating Raper's (1940b) basic experiment and staining the cut fractions with the periodic acid–Schiff method for non-starch poly-saccharides, Bonner, Chiquoine, and Kolderie (1955) showed that the back part of an anterior fraction that previously had the characteristic staining of pre-stalk would slowly change so that ultimately the fraction would have a pre-spore and a pre-stalk zone. The opposite was equally true of a posterior fraction, showing that the cytological peculiarities of pre-stalk and pre-spore are reversible. However this is unfortunately a mere description of the change and reveals nothing concerning the mechanism. A perhaps more hopeful approach is a recent one made by Gregg (1965), who has stained the pseudoplasmodia at various stages with fluorescent antibodies and has also stained segments of migrating pseudoplasmodia of *D. discoideum* at various times after cutting. Surprisingly there is no marked antigenic change in an amputated posterior pre-spore zone, but in the anterior region the cutting produces a great flush of fluorescence, which is similar to what is found at the end of aggregation. Gregg interprets this as meaning that a macromolecule of some juvenile nature is re-synthesized as a result of the operation and is perhaps necessary for the reorganization of this anterior region. It is not clear why this occurs only in the anterior end, although it may reflect the fact that one group of cells has a dominant inhibiting control on the quantity of the other. Also, the fact that this rejuvenescence occurs only at the anterior end may be correlated with Raper's finding that the anterior end takes longer to regulate (Fig. 25).

4. Cell variation

We have described spatial differentiation and examined what little we know of mechanisms, and now we return to the old problem of embryology: Are the differences which arise in this simultaneous differentiation genetic, and therefore fixed and heritable, or are they superficial, environmental fluctuations which are permitted as a result of the repertoire of the genetic constitution? The answer is of course obvious, for the propagation of a colony can occur from a single spore, and therefore one cell possesses the entire genetic endowment for the production of both the stalk and the spore cells. The only challenge to this notion of genetic generality was the hypothesis of the initiator cell, which is now no longer considered tenable.

The idea that the nuclei of an embryo are genetically similar and that spatial differentiation is due to differences in local environments has been the standard conception of embryologists for a considerable number of years. Recently this old scheme has been clothed in new terms, and now the hypothesis is that in the special cytoplasmic environment of a particular region of the developing organism the chromosomal DNA manufactures messenger RNA which goes out to the cytoplasm and, with the help of ribosomal RNA, makes protein. This is a scheme that we all know, and its importance in providing a provisional understanding of a series of key steps is very great. However, it must be remembered that there are other mechanisms which also may be involved in differentiation, such as those which may be concerned with cytoplasmic inheritance.

But in any case it is the localized environment which initiates the change of events, and we will now examine some hypotheses about how this is achieved.

One hypothesis involves the polarity of the cells and has no bearing on the problem of cell variation. As we have discussed, Shaffer (1962, 1964a) has evidence that the cells, at least during aggregation, tend to line up end to end, and he surmises that this relation is retained in the slug. This means that each cell would be like a small arrow, and therefore the properties at the front end of the pseudoplasmodium might well be distinct from those at the posterior end, entirely as a result of cell alignment. Such cell orientation could, for instance, involve the polar movement of substances and therefore serve as a differential mechanism to separate the local environments of the anterior and posterior ends.

Another hypothesis that has been suggested is also connected with polarity, but it does involve cell variation. This is variation of cells with an identical genetic constitution and the possibility that this variation is put to some use by the organism. But before again wandering into the realm of speculation, let us examine all the facts we know concerning variation within the slime molds, for this is solid information that may have uses for future experimental work—my prime interest in writing this book. This discussion of variation will begin with variation among species and strains, proceed to that within a strain, concentrating on mutant types, and finally arrive at the all-important variation within a clone.

One way to study the variation among species and strains is to mix the cells at different stages of development and observe the degree of compatibility. The mixing of cells of different species, particularly at the feeding or vegetative stage, was first done by E. W. Olive (1902). He mixed *D. mucoroides* and *D. purpureum* and found that they produce their characteristic fruiting bodies side by side, showing no coalescence. This matter was thoroughly investigated by Raper and Thom (1941), who not only confirmed Olive's findings, but far extended the observations. They showed that if spores or the vegetative amoebae were mixed, *D. discoideum* and *P. violaceum* would aggregate separately; mixtures of *D. discoideum* and *D. mucoroides* formed common aggregations, but separate fruiting bodies appeared at the center. Grafts made during the migration stage showed some temporary merging in the combination of *D. discoideum* and *D. purpureum*. Furthermore, these two species would form a unified fruiting body when their migrating cell masses were thoroughly intermixed, although spores from this hybrid sorocarp produced, in the F_2 generation, separate sorocarps characteristic of each species.

Gregg (1956) has pursued this matter of adhesion between cells by injecting amoebae of different species into rabbits and obtaining antibodies. He found that the antibodies were species specific on vegetative amoebae, but that this was not the case once the aggregation stage was reached. On the basis of these experiments he suggests that a general tendency for surface adhesion appears at the migration stage, which would possibly explain the observation of Raper and Thom (1941) that the combining of *D. discoideum* and *D. purpureum* cells can only take place at later stages of development. It is, of course, difficult to know to what extent Gregg's experiments reflect the normal properties of cell adhesion, for there is no doubt, as Shaffer (1957a, b) has shown, that stickiness plays an important role in both the aggregation and the later cell association phases. He demonstrated that in combinations of cells of different species there is both a variation in the amount of stickiness, depending upon the combination, and the possibility that species-specific acrasins play an important part in the coalescence of the cells of different species.

If we now turn to the variation among strains, we will find, as has been pointed out on numerous occasions, that each isolate of a cellular slime mold made from nature is likely to show some recognizable morphological characteristic that gives it a distinctive appearance. This is especially evident in the case of *D. mucoroides*, no doubt because it is so frequently isolated. A convenient way of revealing some of the more obvious strain differences is to culture amoebae on a low-nutrient agar, with light coming from one direction. Some strains are small and show little migration, while others will migrate for long periods. Additional differences that have appeared in our cultures are a tendency to form a wavy stalk, tendencies toward periods of stalkless migration, occasional branching, slight differences in PAS staining reactions, presence or absence of macrocysts (Blaskovics and Raper, 1957), a characteristic form to the migrating mass, and a number of more doubtful, less conspicuous ones. These differences are never of sufficient importance to argue for separate species, with the possible exception of size, for *D. minutum* (Raper, 1941a) is perhaps only an extremely small form of *D. mucoroides*.[10]

We have carried out a series of experiments in which the cells of these various strains of *D. mucoroides* are mixed, and find the surprising result that, for the most part, the cells fail to come together, as though they were separate species (Bonner and Adams, 1958). These experiments were extended to include other species as well; in some instances different species did fuse, while many combinations of different strains of the same species did not. Moreover, there were differences in the nature of the combinations, showing varying degrees of fusion and compatibility.

The cells were mixed in two different ways, for the most part with the same results. In one method the aggregating center of one strain was put in among the aggregation streams of another after the center of the latter had been removed. The second method consisted of mixing the cells of two migrating masses very thoroughly with an eyelash.

Depending upon the strains used, the results could be put in one of three categories of compatibility: (1) In the extreme case there is a complete separation of the cells to form

[10] At the upper end of the size scale Singh (1947a) has proposed a *D. giganteum*, but the difficulty is that the size range is almost continuous, in different isolates of *D. mucoroides,* between medium-sized ones and the largest. *D. minutum* is very much smaller than any other *D. mucoroides*, which favors Raper's (1941a) view that it is a separate species.

two separate fruiting bodies. (2) In the intermediate case there is a partial merger; the two discrete cell masses adhere to one another during the migration stage, but they culminate separately, one standing on the sorus of the other. If in these instances the cells are thoroughly mixed, they will regroup during the migration stage so that each strain will be one cohesive mass. (3) The highest degree of compatibility found so far is shown in some masses where there is a single sorus containing two homogeneous, discrete patches of pre-spores, each belonging to one of the strains. Since the cells in this third case do still pull apart and regroup, there is obviously some incompatibility; theoretically it should be possible to find two completely compatible strains in which the spores of both will be spread at random throughout one common sorus.

This matter of sorting out is now a well-recognized phenomenon in experiments with dissociated animal cells. I shall not enter here upon a detailed history of the discovery of the phenomenon; the original idea stems from the work of H. V. Wilson,[11] who pushed sponges through bolting cloth and noted that the dissociated cells reorganized to form new functional sponges. He made the incorrect assumption, at the time of his original experiment, that the differentiated cells had reverted back to some embryonic type and then redifferentiated following coalescence. This error was first corrected by J. S. Huxley,[12] who worked with another species of sponge and showed that the different cells retain their differentiation following dissociation. He pointed out that Driesch's dictum (that the fate of a cell is the function of its position) does not hold in this case; in fact the reverse is true, for the position of a cell is a function of its differen-

[11] *J. Exptl. Zool. 5* (1907) : 245-258.
[12] *Phil. Trans. Roy. Soc. London, Ser. B 202* (1911) : 165-189; *Quart. J. Microscop. Sci. 65* (1921) : 292-321.

tiation. That is, the differentiated cells wander about in the coalesced clump of dissociated cells until they find their proper locations. This point has been confirmed in numerous ways by different workers and is excellently reviewed by P. Brien,[13] who has contributed some evidence himself.

Much the same story can be told of the dissociation of coelenterate cells, and more recently there have been some remarkably convincing experiments on dissociated vertebrate cells. Weiss and Andres[14] injected dissociated presumptive melanoblasts into the blood stream of chick embryos and found that these cells became lodged in their appropriate region in the embryo. Townes and Holtfreter[15] were able to show some specific reorganization in dissociated amphibian embryos, although the best evidence that the cells retain their differentiations comes from more recent work. In particular, Moscona[16] has used an elegant method with a mixture of cells from different species. He used a method he had developed earlier of dissociating cells by the use of trypsin; to mark his cells he used combinations of chick and mouse cells, each of which is histologically recognizable. If mouse and chick cartilage cells are mixed, a mass of continuous cartilage results in which there is a random distribution of mouse and chick cells. However, if mouse cartilage cells are mixed with chick kidney cells, the cells migrate and form discrete masses of cartilage and of kidney tissue. The important proof is that all the cartilage is mouse and all the kidney is chick; there has been no cell transformation but merely a regrouping of the cells.

By far the most impressive explanation of these phenomena comes from the work of Steinberg.[17] He presents evidence

[13] *Arch. Biol. 48* (1937) : 185-268.

[14] *J. Exptl. Zool. 121* (1952) : 449-488.

[15] *J. Exptl. Zool. 128* (1955) : 53-120.

[16] *Proc. Natl. Acad. Sci. U.S. 43* (1957) : 184-194.

[17] Pp. 321-364 in *Cellular Membranes in Development*, M. Locke, ed. Academic Press, New York, 1964.

that fits in with the hypothesis that the sorting out is not the result of specific substances on the surfaces of different cell types but is rather a quantitative effect involving a difference in the amount of cohesive forces between different cell groups. He makes an analogy with the forming of droplets in an emulsion, where the answer to the question of which substance will be continuous and which substance will be surrounded can be predicted entirely on the basis of surface energies. In the experiments on the grafting of different species and strains of slime molds we did not find (as is true with vertebrate tissues) that one cell type surrounds the other; rather, one is always in front or behind the other. The mechanisms producing the two geometrical configurations are perhaps related, and in the case of slime molds the reason one cell type does not surround the other is simply that they are both moving in a polar direction, with the result that one precedes the other.

In the study of variation within a strain, many experiments have been performed to obtain mutants. Extensive work has been done on *D. discoideum, D. mucoroides,* and *D. purpureum* by Sussman and his co-workers, who have produced with ultraviolet irradiation an array of mutant forms (M. Sussman, 1952; R. R. Sussman and M. Sussman, 1953; M. Sussman, 1955a; see also reviews by M. Sussman, 1955b, 1956b, M. Sussman and R. R. Sussman, 1956; Sonneborn, White, and Sussman, 1963). The mutants fall into two general categories: either they stop at some specific stage before culmination (because of a failure to aggregate, for instance), or they culminate with a fruiting body of abnormal character. They can easily be put into one of the following classifications:

1. Aggregateless. The growth is normal but no aggregation takes place. All three species irradiated give aggregateless mutants with considerable frequency; this is the most

commonly observed aberration. There is considerable variation among the particular aggregateless mutants, indicating
that the same change is not induced each time. The type of
medium affects the extent of the expression of the mutant
character; none aggregate on a rich nutrient medium, but if
the amoebae are centrifuged free of bacteria and placed on
washed agar, some of the aggregateless forms proceed to the
beginning of aggregation, and others continue through to
the end (M. Sussman, 1954). Aggregateless mutants were
first reported by Pfützner-Eckert (1950), who observed
them appearing spontaneously in her cultures.

2. Fruitless. These forms undergo aggregation and then
stop, leaving a mound of cells. They appear to be less affected
by the culture conditions and are relatively stable.

3. Bushy. There is normal aggregation with these mutants,
but instead of forming one center, the cell masses break up
into numerous small papillae, each one of which produces a
small, often highly irregular fruiting body. In so doing, they
resemble some of the small forms of cellular slime molds, for
D. lacteum and *D. minutum* will, under favorable culture
conditions, break up into a series of smaller fruiting bodies.
This raises the interesting point that size in the slime molds
can be regulated both by the number of amoebae that enter
an aggregate and by the number of sorocarps formed by an
aggregate. It is conceivable, also, that there is some relation
between this phenomenon and the delayed break-up into small
sorocarps found in *D. polycephalum* and *Polysphondylium*.

4. Glassy. This mutant form is normal up to culmination,
and during culmination it produces a thick structure showing no obvious demarcation between stalk and spore regions.

5. Forked. This is somewhat similar to Glassy, except
that at the tips of the thick, straight stalks there are usually
twin sori produced.

6. Dwarf and Fruity. These mutants are normal morpho-

logically, the only aberration being their minute size. This is especially marked in the mutant Fruity of *D. discoideum,* where an aggregate may consist of very few cells.

There are a number of other mutants listed by M. Sussman (1956b), including some that involve the pigmentation of the spore mass (in *D. discoideum*), and two further aberrations during culmination: Curly and Long-Stemmed (in *D. mucoroides*).

Their most recent mutant of *D. discoideum,* Fr-17, is of considerable interest (Sonneborn, White, and Sussman, 1963). At high population densities it forms flat, irregular cell masses, yet within the mass the cells have differentiated. The interesting thing is that the stalk cells are in central groups, surrounded by loose masses of spores. It would appear as though the polarity of the cells within the mass had been disturbed. At low cell densities fruiting structures do form, but they are quite abnormal in their appearance.

The mutant Fr-17 bears a considerable resemblance to the MV mutant of *D. mucoroides* (Fig. 26) discovered by Filosa (1962), and it is of particular interest that Filosa (1960) found that it could be cured of its abnormality and would resemble the wild type if ethionine was present during the vegetative stage. This approach to differentiation using mutants and chemical cures as well as chemicals producing phenocopies[18] should continue to be a significant one.

Effective use of these mutant strains was made by Sussman in attempting to analyze the steps leading to development. He was able to show that when particular combinations of two mutants that were unable to complete their development (aggregateless and fruitless) were made there was in some cases a synergistic reaction, and development proceeded farther than would have been the case normally for

[18] This has been done especially with the use of amino acid analogues (Kostellow, 1956; Filosa, 1960).

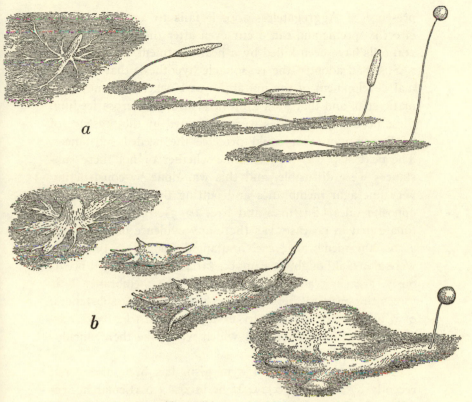

Fig. 26. a) *D. mucoroides*; wild type. b) MV mutant of *D. mucoroides*. (Drawing by J. L. Howard, courtesy of the *Scientific American*.)

either one alone (M. Sussman, 1954). For example, it is possible to mix Fruitless-1, which normally forms loose aggregates, with Aggregateless-53, which never aggregates, and the result will be normal, mature fruiting bodies. He has found a dozen or so cases of such synergism, and curiously enough, one case of antagonism.[19] The stock Aggregateless-208 can aggregate on washed agar but not on the normal nutrient medium. However, on washed agar, in the

[19] This kind of antagonism was first reported by Pfützner-Eckert (1950).

presence of Aggregateless-206, it fails to aggregate. This effect is specific and can occur even after the Aggregateless-206 cells have been killed by a heat treatment.

Sussman adopted the reasonable hypothesis that the normal development of the mutants was blocked at some enzymatic step, and that by a kind of syntropy or cross feeding, these necessary syntheses were by-passed by the presence of another mutant that could produce the needed substances. The next step was to determine whether or not these substances were diffusible, and this was done by constructing very fine agar membranes and putting the two mutants on opposite sides (Sussman and Lee, 1955). It is most unfortunate that in no case was there any evidence of synergism across the membrane; each mutant developed as though it were alone, although, as Runyon (1942) had reported previously, acrasin readily passes through the membrane. This forces Sussman to the conclusion that these key substances, of which we are so eager to learn more, must be passed directly by cell contact, a fact which will make their elucidation more difficult.

A most curious case of synergism has been discovered recently by Kahn (1964a). If he mixes a particular aggregateless mutant of *D. purpureum* with the wild type, all the aggregateless cells enter the normal fruiting bodies, even if the aggregateless cells are greatly in excess of the normal ones. But the most remarkable aspect of the case is that all the spores give rise to wild-type progeny; not only does the wild type cure the aggregateless cells during the immediate generation, but the cure is permanent and the mutant cells are completely converted to stable wild-type cells.

Another example of synergism, which might be compared to the gene dosage of the geneticist, was found by Filosa (1962). In this example there are also mixtures of cell types, and not only is the phenotypic expression affected, but also there are interesting control or balance mechanisms that maintain a relatively constant proportion in the number of the cells of the different types.

His work began with the discovery that any one strain or isolate of a cellular slime mold (which has been repeatedly recultured by the mass transfer of spores) is not usually made up of one genetic cell strain, but of two or more. That is, if he makes careful, single-spore platings from the fruiting bodies of a particular isolate, the majority of the spores give rise to colonies identical in appearance with the parent, but a certain per cent of the spores produce colonies that are distinctly different. This point is implicit in an experiment of Blaskovics and Raper (1957), where they show that the macrocyst-producing character is different in different clones of a particular strain of *D. mucoroides*; further, Sussman (1956b) makes the brief remark that mutants have been found in unirradiated populations of *D. mucoroides*.

In the case of a strain of *D. mucoroides* that has been transferred repeatedly by mass inoculation, Filosa found that in single-spore isolates roughly 87 per cent resemble the parent, and 13 per cent have distinctly different appearances which fall into three categories: (1) The most common (which averages approximately 9 per cent) is a form in which the migration period is without a stalk (the MV mutant, Fig. 26). It does not really resemble *D. discoideum*, for the cell masses are messy and irregular, frequently disintegrating and breaking up into separate amoebae. Also, it has no basal disc, and histological preparations of the migrating stage show a few abortive stalk cells in the tip region of the abnormally pointed migrating mass. (2) Occasionally one finds clones that differ from the parent stock in that they have a reduced period of migration and a peculiar aggregation pattern, which is somehow slowed down so that aggregation proceeds well into culmination, and there is a

rope of cells leading up the stalk to the rising cell mass.
(3) Finally, aggregateless mutants appear in less than 1 per
cent of the cells.

Some of these changes are very stable, for in the case of
the MV mutant, Filosa found that the proportion of 9 per
cent is very roughly retained, at least for 25 generations.
However, in other cases new mutants will appear; in a clone
derived from one "normal" or "wild type" spore of *D.
mucoroides*, after 20 generations of mass spore inoculation
the spores in a sorus were found to consist of almost 50 per
cent of a new mutant. This was especially interesting in
that the phenotype showed no indication of the mutant form,
despite the fact that half the cells were of abnormal genotype.

To determine the effect of the stalkless migration clone
(MV) on the normal cells of strain No. 11 of *D. mucoroides*,
Filosa mixed the two kinds of cells in different proportions
and at various stages; he mixed spores, centrifuged vegetative amoebae, and made grafts at the aggregation stage.
In this way he could vary at will the ratio of MV to normal
cells and even test the resultant ratio in the sorus of the mixed
sorocarp. In brief, if he provides 50 per cent or more MV
cells, the fruiting bodies will be normal in appearance, showing spore ratios which unexpectedly will be predominately
MV (on the order of 6 to 1 in favor of MV). This might indicate that the normal cells moved to the anterior end and became trapped in the stalk. It is of interest that Filosa showed
that if a migrating mass is made by grafting an MV tip into
the aggregation streams of normal cells (or vice versa), the
result is a normal fruiting body. In other words, if there are
large numbers of cells of each cell type, again the phenotypic
effect of the normal cells is expressed.

It is impossible to know at the moment whether the balance
of cell numbers of different strains (which has its parallel in
the balance of genetically different haploid nuclei in a hetero-

caryon of *Neurospora*) is mediated by a mechanism similar to the balance between stalk and spore cells. Furthermore we cannot know how phenotypic expression is achieved by a minority cell type. These questions would seem to be important in our consideration of differentiation, but at the moment they do little more than tantalize the imagination. We come, therefore, to the final problem of cell variation, which has a more direct and immediate relation to differentiation. This is the problem of variation within a clone.

Skupienski (1920) was the first to isolate a single spore (in *D. mucoroides*) and show that normal fruiting bodies would result after a period of growth. This has been confirmed by Raper (1951) and myself (Bonner, 1952) for *D. discoideum*, and Sussman (1951) did an extensive series of experiments substantiating the notion. He also showed that single vegetative cells are capable of giving rise to whole colonies, although his data for the cells at the later stages of association are less convincing, for it is not clear that he did in fact separate the sticky masses of cells. But since aggregating and migrating cells may be made to revert back to vegetative cells, there is no reason to doubt that any cell, except those that have been trapped in the stalk for some time, has the power of starting a new colony upon isolation. The reason for arguing that the stalk cells lose this capacity is that there is good evidence that cells in the stalk slowly degenerate and die.[20]

If a sorocarp may arise from a single cell, we are faced with the problem of how this one cell produces both stalk and spore cells. Furthermore, it does so by first producing a large number of independent cells that then stream together into masses. The first question that must be answered is whether or not these separate vegetative cells, just prior to aggregation, are totipotent (or dipotent in this case, since they can

[20] See page 67.

only become spores or stalk cells). The other possibility is that they have already become segregated in their character, and that they assemble in aggregation according to their future role. These are not, however, mutually exclusive notions as we shall suggest presently.

The fact that the cells retain their dual potencies up to the last minute before differentiation is strongly supported by the studies on regulation following transection of the pseudo-plasmodium which have just been discussed.

The notion that there may be some sorting out or shuffling of cells during the early development of the cell mass also has considerable support, although we do not yet know whether or not this is of significance to the process of differentiation. The evidence consists of two kinds: the fact that the individual cells are not identical but differ significantly, and the fact that the cells are shuffled in development.

The kind of cell variation that can be most easily measured is the variation in cell size. The size range of both cells and nuclei in the different stages of development is unusually wide (Bonner and Frascella, 1953; Bonner, Chiquoine, and Kolderie, 1955). This may be seen especially dramatically if the size range of *Dictyostelium* cells is compared with the range of different unicellular organism as listed by Adolph;[21] there is little that approaches the variability of the cells of the slime mold.

Jennings[22] showed that if protozoans from a clone were selected for size, that is, if the smallest or the largest were isolated, and their progeny measured, there was in each case the same size frequency distribution in the progeny as in the original clone. We have done the same thing with cellular

[21] In *The Regulation of Size as Illustrated in Unicellular Organisms.* C. C. Thomas, Baltimore, 1931.

[22] Chapter 15 in *Protozoa in Biological Research*, G. N. Calkins and F. M. Summers, eds., Columbia University Press, 1941. See also J. E. Ackert. *Genetics 1* (1916): 387-405.

slime mold cells, using a strain of *D. mucoroides*; we measured the frequency distribution of spore size for 100 spores. Then single spores were isolated and measured, after which each spore was allowed to produce an F_1 generation.[23] The size range of spores produced by this generation was measured, and in no case did the range differ significantly from the range of the parental strains, even when the original spore of the clone was on the large or the small end of the range. Admittedly, these experiments would be even more convincing if this procedure were repeated over a series of generations, but the experiments as they now stand are sufficient to justify the statement that the size of the parent cell does not affect either the mean size or the variability of its progeny; the variation in size within a clone is independent of the size of the parent cell. What is inherited is the ability to have a range of size variation, and this is not determined by the size of any one cell. This kind of inheritance has been discussed in detail elsewhere and referred to as *range variation*.[24]

Another clear-cut range variation in the cell population is that of speed of movement. It is an easy matter to show that the aggregating cells vary greatly in their rate of movement, and evidence has been given that all cells do not move at the same speed during the migration stage.[25] Also, it will be recalled, it was possible to show, by grafting vitally colored

[23] This was done by using modifications of a technique of Lederberg (*J. Bacteriol. 68*: 258-259, 1954). The spores in suspension were drawn up into a small pipette and dispensed in minute drops onto a slide. The slide was marked off in squares and covered with a layer of heavy mineral oil. Then each drop was examined with a microscope for those with only one spore. The spore was measured and transferred to an environment favorable for growth.

[24] Bonner, J. T., *Size and Cycle*, Princeton University Press, 1965.

[25] As was mentioned previously (page 98), we have no evidence on whether this shifting of positions of cells within the mass is due to differences in velocity or whether some other factor might be operating, such as selective adhesiveness.

anterior portions into colorless posterior portions, that the colored cells moved anteriorly as a group, while posterior cells placed anteriorly lagged, finally settling at the hind end (Bonner, 1952). This suggested that either before or during migration there was a sorting out of fast and slow cells, the fast ones populating the front end, the slow ones, the hind end. Subsequent work has indicated that this major sorting out takes place between aggregation and migration (as will be discussed presently) and that during migration there are only a few especially fast and especially slow cells that continue to change their relative positions (Bonner, 1957; Bonner and Adams, 1958).

Experiments have also been performed to determine whether or not this difference in velocity is inherited (Bonner, 1952). Anterior ends of migrating slugs of *D. discoideum*, which presumably contained all fast cells, were cut off and allowed to fruit. The spores of these fruiting bodies were sown on fresh media, and a migrating slug of the next generation was again cut so that its anterior end fruited. This was repeated for ten generations, but even after this continued selection the fruiting bodies that arose from this long line of especially fast spores showed no differences in their speed of migration, or in any other character, from a normal population of *D. discoideum*. This experiment was also carried out using the posterior end, with the same results. Therefore speed is not affected by the speed of the parent cells; what is passed on by any one cell, fast or slow, is the ability to produce cells of different rates of movement—another example of range variation.

These characteristics, speed and size, are rather superficial, and unfortunately we do not have—what would be more interesting—examples of biochemical range variation, although it is possible in histological and histochemical preparations to see striking differences between cells. The fact that the cells within one pseudoplasmodium do differ even when they are members of the same clone is nevertheless unequivocal. Our next question is whether these different cells are shuffled or sorted out in any organized fashion or whether they always remain in a random condition. This has been partially answered above by the observations on fast and slow cells, but there is even better evidence to support the same point. With the use of vital dyes and mutant markers it is possible to show that between aggregation and migration there is a major rearrangement of cells (Bonner, 1959a). This is a period of violent internal reorganization.

The only experiments which bear on the question of whether or not this reshuffling has a relation to differentiation are those of Takeuchi (1963). He obtained spore antiserum by injecting into rabbits spores of *D. mucoroides* separated from stalks. This spore antiserum was conjugated with a fluorescent dye, and various stages of development were examined to detect spore protein (or antigen). It was found to be present in some but not all of the cells before aggregation; during aggregation these cells were randomly distributed. It was only after aggregation was completed that all the cells with spore protein appeared in the posterior prespore region, as though they had been sorted out. Unfortunately, as Takeuchi himself points out, along with Gregg (1965), there are other possible interpretations of these results; for instance, cells in the anterior region could destroy their spore antigens, while those in the posterior end could synthesize them in abundance.

Here we have again the question of whether the fate of a cell is the function of its position or whether the position of a cell is determined by its fate or inner differentiation. The confusing thing is that, if one looks at all the facts we have presented, there would seem to be evidence for both possibilities. There is clearly range variation and reshuffling, and at the same time there is clear evidence that the cells are totipotent and capable of proportionate regulation.

This apparent paradox can be solved in two ways. The evidence concerning totipotency is unshakable, and all hypotheses must therefore admit this fact, but we could assume that the range variation and the reshuffling simply have nothing to do with differentiation. In this case one would assume that all the differences between the front and the hind end of the cell mass arise as a result of some such agency as the summation of all the individual cell polarities. By postulating that the cells in the mass all point the same way, and by assuming that this means the movement of substance in a polar fashion, one could argue for a gradient of chemical differences between the front and hind end, and even a system whereby exact proportions between spore and stalk could be maintained.

The other possibility is that the range variation and the shuffling of the cells contribute to the establishment of a gradient that leads to proportional development. In this hypothesis, sorting out (along with polar orientation) results in a graded mass of cells at the beginning of migration. This gradient produces the different environments required for spore and stalk differentiation, and there must be some sort of communication between these two groups of cells which maintains the exact proportions.

Speculation is easy, but it is facts that we need. At this stage in our knowledge of these organisms we have little hope that any particular hypothesis will last long, for knowledge is advancing rapidly. The only purpose of ending this book with guesses is to show what kinds of information we want, what kinds of facts we need, if we wish ultimately to understand how cellular slime molds develop.

A Bibliography of the Cellular Slime Molds

This is a complete list of books and articles which have made a significant contribution to the biology of the Acrasiales up through 1965 (with some references in 1966). Abstracts and theses are not included unless all or part of the material has not appeared in a printed article.

Agnihothrudu, V. (1956). Occurrence of Dictyosteliaceae in the rhizosphere of plants in Southern India. *Experentia 12*: 149-150.

Allderdice, J. D. (1965). A growth promoting substance in *E. coli*. Senior thesis, Princeton University.

Allen, J. R., S. H. Hutner, E. Goldstone, J. J. Lee, and M. Sussman (1963). Culture of the acrasian *Polyspondylium pallidum,* WS-320, in defined media. *J. Protozool. 10 (Suppl)*: 13.

Arnaud, G. (1948). Les Heimerliacées, subdivision des Acrasiales? *Compt. Rend. Acad. Sci. 266*: 1744-1746.

Arnaud, G. (1949). Les Heimerliacées, subdivision des Acrasiales? *Botaniste 34*: 35-36.

Arndt, A. (1937). Untersuchungen über *Dictyostelium mucoroides* Brefeld. *Roux' Arch. Entwicklungsmech. Organ. 136*: 681-747.

Belousov, L. V., E. B. Vsevolodov, and V. A. Golichenkov (1963). Razvitie slizistykh gribov i nekotorye problemy eksperimental' noi embriologii. [The development of slime molds and some problems of experimental embryology.] *Usp. Sovrem. Biol. 55*: 109-117.

Blaskovics, J. C. and K. B. Raper (1957). Encystment stages of *Dictyostelium. Biol. Bull. 113*: 58-88.

Bonner, J. T. (1944). A descriptive study of the development of the slime mold *Dictyostelium discoideum. Am. J. Botany 31*: 175-182.

Bonner, J. T. (1947). Evidence for the formation of cell aggregates by chemotaxis in the development of the slime mold *Dictyostelium discoideum. J. Exptl. Zool. 106*: 1-26.

Bonner, J. T. (1949). The demonstration of acrasin in the later stages of the development of the slime mold *Dictyostelium discoideum. J. Exptl. Zool. 110*: 259-271.

Bonner, J. T. (1950). Observations on polarity in the slime mold *Dictyostelium discoideum. Biol. Bull. 99*: 143-151.

Bonner, J. T. (1952). The pattern of differentiation in amoeboid slime molds. *Am. Naturalist 86*: 79-89.

Bonner, J. T. (1957). A theory of the control of differentiation in the cellular slime molds. *Quart. Rev. Biol. 32*: 232-246.

Bonner, J. T. (1958). *The Evolution of Development.* Cambridge University Press.

Bonner, J. T. (1959a). Evidence for the sorting out of cells in the development of the cellular slime molds. *Proc. Natl. Acad. Sci. U. S. 45*: 379-384.

Bonner, J. T. (1959b). *The Cellular Slime Molds,* 1st edn. Princeton University Press.

Bonner, J. T. (1959c). Differentiation of social amoebae. *Sci. Am. 201* (December) : 152-162.

Bonner, J. T. (1960). Development in the cellular slime molds: the role of cell division, cell size, and cell number. Pp. 3-20 in *Developing Cell Systems and Their Control,* 18th Growth Symposium, D. Rudnick, ed. Ronald, New York.

Bonner, J. T. (1963a). How slime molds communicate. *Sci. Am. 209* (August) : 84-93.

Bonner, J. T. (1963b). Epigenetic development in the cellular slime moulds. *Symp. Soc. Exptl. Biol. 17*: 341-358.

Bonner, J. T. (1965). Physiology of development in cellular slime molds. (Acrasiales). Pp. 612-640 in *Encyclopedia of Plant Physiology*, Vol. 15/1, W. Ruhland, ed.

Bonner, J. T. and M. S. Adams (1958). Cell mixtures of different species and strains of cellular slime moulds. *J. Embryol. Exptl. Morphol. 6*: 346-356.

Bonner, J. T., A. D. Chiquoine, and M. Q. Kolderie (1955). A histochemical study of differentiation in the cellular slime molds. *J. Exptl. Zool. 130*: 133-158.

Bonner, J. T., W. W. Clarke, Jr., C. L. Neely, Jr., and M. K. Slifkin (1950). The orientation to light and the extremely sensitive orientation to temperature gradients in the slime mold *Dictyostelium discoideum. J. Cellular Comp. Physiol. 36*: 149-158.

Bonner, J. T. and M. R. Dodd (1962a). Aggregation territories in the cellular slime molds. *Biol. Bull. 122*: 13-24.

Bonner, J. T. and M. R. Dodd (1962b). Evidence for gas-induced orientation in the cellular slime molds. *Develop. Biol. 5*: 344-361.

Bonner, J. T. and D. Eldredge, Jr. (1945). A note on the rate of morphogenetic movement in the slime mold, *Dictyostelium discoideum. Growth 9*: 287-297.

Bonner, J. T. and E. B. Frascella (1952). Mitotic activity in relation to differentiation in the slime mold *Dictyostelium discoideum. J. Exptl. Zool. 121*: 561-571.

Bonner, J. T. and E. B. Frascella (1953). Variations in cell size during the development of the slime mold, *Dictyostelium discoideum. Biol. Bull. 104*: 297-300.

Bonner, J. T. and M. E. Hoffman (1963). Evidence for a substance responsible for the spacing pattern of aggregation and fruiting in the cellular slime molds. *J. Embryol. Exptl. Morphol. 11*: 571-589.

Bonner, J. T., A. P. Kelso, and R. G. Gillmor (1966). A new approach to the problem of aggregation in the cellular slime molds. *Biol. Bull. 130*: 28-42.

Bonner, J. T., P. G. Koontz, Jr., and D. Paton (1953). Size in relation to the rate of migration in the slime mold *Dictyostelium discoideum. Mycologia 45*: 235-240.

Bonner, J. T. and M. J. Shaw (1957). The role of humidity in the differentiation of the cellular slime molds. *J. Cellular Comp. Physiol. 50*: 145-154.

Bonner, J. T. and M. K. Slifkin (1949). A study of the control of differentiation: the proportions of stalk and spore cells in the slime mold *Dictyostelium discoideum. Am. J. Botany 36*: 727-734.

Bonner, J. T. and F. E. Whitfield (1965). The relation of sorocarp size to phototaxis in the cellular slime mold *Dictyostelium purpureum. Biol. Bull. 128*: 51-57.

Bradley, S. G. and M. Sussman (1952). Growth of ameboid slime molds in one-membered cultures. *Arch. Biochem. Biophys. 39*: 462-463.

Bradley, S. G. and M. Sussman (1954). Physiology of the aggregation stage in the development of the cellular slime molds (Abstract). *Bacteriol. Proc.*, p. 32.

BIBLIOGRAPHY

Bradley, S. G., M. Sussman, and H. L. Ennis (1956). Environmental factors affecting the aggregation of the cellular slime mold, *Dictyostelium discoideum. J. Protozool. 3*: 33-38.

Brefeld, O. (1869). *Dictyostelium mucoroides*. Ein neuer Organismus aus der Verwandtschaft der Myxomyceten. *Abhandl. Senckenberg. Naturforsch. Ges. 7*: 85-107.

Brefeld, O. (1884). *Polysphondylium violaceum* und *Dictyostelium mucoroides* nebst Bemerkungen zur Systematik der Schleimpilze. *Untersuchungen aus dem Gesammtgebiet der Mykologie 6*: 1-34.

Brühmüller, M. and B. E. Wright (1963). Glutamate oxidation in the differentiating slime mold. II. Studies *in vitro. Biochim. Biophys. Acta 71*: 50-57.

Bulkley, G. B. (1965). A study of differentiation in the cellular slime mold *Dictyostelium discoideum* (Raper). Senior thesis, Princeton University.

Cavender, J. C. and K. B. Raper (1965a). The Acrasieae in nature. I. Isolation. *Am. J. Botany 52*: 294-296.

Cavender, J. C. and K. B. Raper (1965b). The Acrasieae in nature. II. Forest soil as a primary habitat. *Am. J. Botany 52*: 297-302.

Cavender, J. C. and K. B. Raper (1965c). The Acrasieae in nature. III. Occurrence and distribution in forests of Eastern North America. *Am. J. Botany 52*: 302-308.

Ceccarini, C. (1966). The biochemical relationship between trehalase and trehalose during growth and differentiation in the cellular slime mold, *Dictyostelium discoideum*. Ph.D. thesis, Princeton University.

Ceccarini, C. and M. Filosa (1965). Carbohydrate content during development of the slime mold, *Dictyostelium discoideum. J. Cellular Comp. Physiol. 66*: 135-140.

Chatton, E. (1912). Entamibe (*Loeschia* sp.) et myxomycète (*Dictyostelium mucoroides* Brefeld) d'un singe. *Bull. Soc. Pathol. Exotique 5*: 180-184.

Cienkowsky, L. (1873). *Guttulina rosea*. Trans. bot. section 4th meeting Russian naturalists at Kazan (in Russian).

Clegg, J. S. and M. F. Filosa (1961). Trehalose in the cellular slime mould *Dictyostelium mucoroides. Nature 192*: 1077-1078.

Coemans, E. (1863). Recherches sur le polymorphisme et les différents appareils de reproduction chez les mucorinées. *Bull.*

BIBLIOGRAPHY

Acad. Roy. Sci. Lettres Beaux Arts Belg. 16 (2ᵉ series):
177-188.

Cohen, A. L. (1953a). The effect of ammonia on morphogenesis in the Acrasieae. *Proc. Natl. Acad. Sci. U. S. 39*: 68-74.

Cohen, A. L. (1953b). The isolation and culture of opsimorphic organisms. I. Occurrence and isolation of opsimorphic organisms from soil and culture of Acrasieae on a standard medium. *Ann. N. Y. Acad. Sci. 56*: 938-943.

Cohen, A. L. (1965). Slime molds. *Encyclopaedia Britannica 20*: 797-798.

Cook, W. R. I. (1939). Some observations on *Sappinia pedata* Dang. *Trans. Brit. Mycol. Soc. 22*: 302-306.

Dangeard, P. A. (1896). Contribution à l'étude des Acrasiées. *Botaniste 5*: 1-20.

Davidoff, F. and E. D. Korn (1962). Lipids of *Dictyostelium discoideum*: Phospholipid composition and the presence of two new fatty acids, cis,cis-5,11-octadecadienoic and cis,cis-5, 9-hexadecadienoic acids. *Biochem. Biophys. Res. Commun. 9*: 54-58.

Davidoff, F. and E. D. Korn (1963a). Fatty acid and phospholipid composition of the cellular slime mold, *Dictyostelium discoideum*: the occurrence of previously undescribed fatty acids. *J. Biol. Chem., 238*: 3199-3209.

Davidoff, F. and E. D. Korn (1963b). The biosynthesis of fatty acids in the cellular slime mold, *Dictyostelium discoideum*. *J. Biol. Chem. 238*: 3210-3215.

Dehaan, R. L. (1959). The effects of the chelating agent ethylenediamine tetra-acetic acid on cell adhesion in the slime mould *Dictyostelium discoideum*. *J. Embryol. Exptl. Morphol. 7*: 335-343.

Dutta, S. K. and E. D. Garber (1961). The identification of physiological races of a fungal phytopathogen using strains of the slime mold *Acrasis Rosea*. *Proc. Natl. Acad. Sci. U. S. 47*: 990-993.

Ennis, H. L. and M. Sussman (1958a). Synergistic morphogenesis by mixtures of *Dictyostelium discoideum* wild-type and aggregateless mutants. *J. Gen. Microbiol. 18*: 433-449.

Ennis, H. L. and M. Sussman (1958b). The initiator cell for slime mold aggregation. *Proc. Natl. Acad. Sci. U. S. 44*: 401-411.

Ennis, H. L. and M. Sussman (1958c). The initiator cell for

slime mold aggregation (Abstract). *Bacteriol. Proc.*, p. 32.
Faust, R. G. and M. F. Filosa (1959). Permeability studies on the amoebae of the slime mold, *Dictyostelium mucoroides*. *J. Cellular Comp. Physiol. 54*: 297-298.
Fayod, V. (1883). Beitrag zur Kenntnis niederer Myxomyceten. *Botan. Zeitung 41*: 169-177.
Feiro, A. D. (1961). Presenting—The cellular slime molds. *Am. Biol. Teacher 23*: 501-505.
Filosa, M. F. (1960). The effects of ethionine on the morphogenesis of cellular slime molds (Abstract). *Anat. Record 138*: 348.
Filosa, M. F. (1962). Heterocytosis in cellular slime molds. *Am. Naturalist 96*: 79-91.
Francis, D. W. (1959). Pseudoplasmodial movement in *Dictyostelium discoideum*. M.S. thesis, University of Wisconsin.
Francis, D. W. (1962). The movement of pseudoplasmodia of *Dictyostelium discoideum*. Ph.D. thesis, University of Wisconsin.
Francis, D. W. (1964). Some studies on phototaxis of *Dictyostelium*. *J. Cellular Comp. Physiol. 64*: 131-138.
Francis, D. W. (1965). Acrasin and the development of *Polysphondylium pallidum*. *Develop. Biol. 12*: 329-346.
Fuller, M. S. and R. M. Rakatansky (1966). A preliminary study of the corotenoids in *Acrasis rosea*. *Can. J. Botany 44*: 269-274.
Gamble, W. J. (1953). Orientation of the slime mold *Dictyostelium discoideum* to light. Senior thesis, Princeton University.
Gerisch, G. (1959). Ein Submerskulturverfahren für entwicklungsphysiologische Untersuchungen an *Dictyostelium discoideum*. *Naturwiss. 46*: 654-656.
Gerisch, G. (1960). Zellfunktionen und Zellfunktionswechsel in der Entwicklung von *Dictyostelium discoideum*. II. Agglutination und Induktion der Fruchtkörperpolarität. *Roux' Arch. Entwicklungsmech. Organ. 152*: 632-654.
Gerisch, G. (1961a) Zellfunktionen und Zellfunktionswechsel in der Entwicklung von *Dictyostelium discoideum*. II. Aggregation homogener Zellpopulationen und Zentrenbildung. *Develop. Biol. 3*: 685-724.
Gerisch, G. (1961b). Zellfunktionen und Zellfunktionswechsel in der Entwicklung von *Dictyostelium discoideum*. III. Ge-

trennte Beeinflussung von Zelldifferenzierung und Morphogenese. *Roux' Arch. Entwicklungsmech. Organ. 153*: 158-167.

Gerisch, G. (1961c). Zellfunktionen und Zellfunktionswechsel in der Entwicklung von *Dictyostelium discoideum*. V. Stadienspezifische Zellkontaktbildung und ihre quantitative Erfassung. *Exptl. Cell Res. 25*: 535-554.

Gerisch, G. (1961d). Zellkontaktbildung vegetativer und aggregationsreifer Zellen von *Dictyostelium discoideum*. *Naturwiss. 48*: 436-437.

Gerisch, G. (1962a). Die Zellulären Schleimpilze als Objekte der Entwicklungsphysiologie. *Ber. Deut. Botan. Ges. 75*: 82-89.

Gerisch, G. (1962b). Zellfunktionen und Zellfunktionswechsel in der Entwicklung von *Dictyostelium discoideum*. IV. Der Zeitplan der Entwicklung. *Roux' Arch. Entwicklungsmech. Organ. 153*: 603-620.

Gerisch, G. (1962c). Zellfunktionen und Zellfunktionswechsel in der Entwicklung von *Dictyostelium discoideum*. VI. Inhibitoren der Aggregation, ihr Einfluss auf Zellkontaktbildung und morphogenetische Bewegung. *Exptl. Cell Res. 26*: 462-484.

Gerisch, G. (1963). Eine für *Dictyostelium* ungewöhnliche Aggregationsweise. *Naturwiss. 50*: 160-161.

Gerisch, G. (1964a). Entwicklung von *Dictyostelium,* (Commentary on a film). *Publikationen zu Wissenschaftlichen Filmen, 1*: 127-140.

Gerisch, G. (1964b). Die Bildung des Zellverbandes bei *Dictyostelium minutum*. I. Übersicht über die Aggregation und den Funktionswechsel der Zellen. *Roux' Arch. Entwicklungsmech. Organ. 155*: 342-357.

Gerisch, G. (1965a) *Dictyostelium purpureum* (Acrasina). Vermehrungsphase. *Publikationen zu Wissenschaftlichen Filmen 1*: 237-244.

Gerisch, G. (1965b). *Dictyostelium purpureum* (Acrasina). Aggregation und Bildung des Sporophors. *Publikationen zu Wissenschaftlichen Filmen 1*: 245-254.

Gerisch, G. (1965c). *Dictyostelium discoideum* (Acrasina). Aggregation und Bildung des Sporophors. *Publikationen zu Wissenschaftlichen Filmen 1*: 255-264.

Gerisch, G. (1965d). *Dictyostelium minutum* (Acrasina). Aggregation. *Publikationen zu Wissenschaftlichen Filmen* 1: 265-278.

Gerisch, G. (1965e). Eine Mutante von *Dictyostelium minutum* mit blockierter Zentrengrundung. *Z. Naturforsch. 20b*: 298-301.

Gerisch, G. (1965f). Stadienspezifische Aggregationsmuster von *Dictyostelium discoideum. Roux' Arch. Entwicklungsmech. Organ. 156*: 127-144.

Gerisch, G. (1965g). Spezifische Zellkontakte als Mechanismen der tierischen Entwicklung. *Umschau 65*: 392-395.

Gerisch, G. (1966). Die Bildung des Zellverbandes bei *Dictyostelium minutum*. II. Analyse der Zentrengrundung anhand von Filmaufnahmen. *Roux' Arch. Entwicklungsmech. Organ. 157*: 174-189.

Gezelius, K. (1959). The ultrastucture of cells and cellulose membranes in Acrasiae. *Exptl. Cell Res. 18*: 425-453.

Gezelius, K. (1961). Further studies in the ultrastructure of *Acrasiae. Exptl. Cell Res. 23*: 300-310.

Gezelius, K. (1962). Growth of the cellular slime mold *Dictyostelium discoideum* on dead bacteria in liquid media. *Physiol. Plantarum 15*: 587-592.

Gezelius, K. and B. G. Rånby (1957). Morphology and fine structure of the slime mold *Dictyostelium discoideum. Exptl. Cell Res. 12*: 265-289.

Gezelius, K. and B. E. Wright (1965). Alkaline phosphatase in *Dictyostelium discoideum. J. Gen. Microbiol. 38*: 309-327.

Gregg, J. H. (1950). Oxygen utilization in relation to growth and morphogenesis of the slime mold *Dictyostelium discoideum. J. Exptl. Zool. 114*: 173-196.

Gregg, J. H. (1956). Serological investigations of cell adhesion in the slime molds, *Dictyostelium discoideum, D. purpureum,* and *Polysphondylium violaceum. J. Gen. Physiol. 39*: 813-820.

Gregg, J. H. (1957). Serological investigations of aggregateless variants of the slime mold, *Dictyostelium discoideum* (Abstract). *Anat. Record 128*: 558.

Gregg, J. H. (1960). Surface antigen dynamics in the slime mold, *Dictyostelium discoideum. Biol. Bull. 118*: 70-78.

Gregg, J. H. (1961). An immunoelectrophoretic study of the slime mold *Dictyostelium discoideum. Develop. Biol. 3*: 757-766.

Gregg, J. H. (1964). Developmental processes in cellular slime molds. *Physiol. Rev. 44*: 631-656.

Gregg, J. H. (1965). Regulation in the cellular slime molds. *Develop. Biol. 12*: 377-393.

Gregg, J. H. (1966a). Organization and synthesis in the cellular slime molds. Pp. 235-281 in *The Fungi, An Advanced Treatise,* Vol. 2, G. C. Ainsworth and A. S. Sussman, eds. Academic Press, Inc., New York.

Gregg, J. H. (1966b). Manipulating cellular slime molds. In *Techniques for the Study of Development,* F. H. Wilt and N. K. Wessells, eds. T. Y. Crowell Co., New York. (In press.)

Gregg, J. H. (1966c). Antigen synthesis during reorganization in the cellular slime molds. In *The Molecular Aspects of Development,* R. A. Deering, ed. N.A.S.A. Publication. (In press.)

Gregg, J. H. and R. D. Bronsweig. (1954). The carbohydrate metabolism of the slime mold *Dictyostelium discoideum,* during development (Abstract). *Biol. Bull. 107*: 312.

Gregg, J. H. and R. D. Bronsweig (1956a). Dry weight loss during culmination of the slime mold, *Dictyostelium discoideum. J. Cellular Comp. Physiol. 47*: 483-488.

Gregg, J. H. and R. D. Bronsweig (1956b). Biochemical events accompanying stalk formation in the slime mold, *Dictyostelium discoideum. J. Cellular Comp. Physiol. 48*: 293-300.

Gregg, J. H., A. L. Hackney, and J. O. Krivanek (1954). Nitrogen metabolism of the slime mold *Dictyostelium discoideum* during growth and morphogenesis. *Biol. Bull. 107*: 226-235.

Gregg, J. H. and C. W. Trygstad (1958). Surface antigen defects contributing to developmental failure in aggregateless variants of the slime mold, *Dictyostelium discoideum. Exptl. Cell Res. 15*: 358-369.

Grimm, M. (1895). Ueber den Bau und die Entwickelungsgeschichte von *Dictyostelium mucoroides* Bref. (Résumé). *Scripta Botan. Hort. Univ. Imp. Petersburg 4*: 279-298.

Harper, R. A. (1926). Morphogenesis in *Dictyostelium. Bull. Torrey Botan. Club 53*: 229-268.

Harper, R. A. (1929). Morphogenesis in *Polysphondylium. Bull. Torrey Botan. Club 56*: 227-258.

Harper, R. A. (1932). Organization and light relations in *Polysphondylium. Bull. Torrey Botan. Club 59*: 49-84.

Heftmann, E., B. E. Wright, and G. U. Liddel (1959). Identification of a sterol with acrasin activity in a slime mold. *J. Am. Chem. Soc. 81*: 6525.

Heftmann, E., B. E. Wright, and G. U. Liddel (1960). The isolation of delta²²-stigmasten-3 beta-ol from *Dictyostelium discoideum. Arch. Biochem. Biophys. 91*: 266-270.

Heller, S. A. and M. C. Miles (1961). The effect of humidity and light on sorocarp density in *Dictyostelium purpureum*. Senior thesis, Princeton University.

Hirschberg, E. (1955). Some contributions of microbiology to cancer research. *Bacteriol. Rev. 19*: 65-78.

Hirschberg, E. and G. Merson (1955). Effect of test compounds on the aggregation and culmination of the slime mold *Dictyostelium discoideum. Cancer Res. Suppl. 3*: 76-79.

Hirschberg, E. and H. P. Rusch (1950). Effects of compounds of varied biochemical action on the aggregation of a slime mold, *Dictyostelium discoideum. J. Cellular Comp. Physiol. 36*: 105-113.

Hirschberg, E. and H. P. Rusch (1951). Effect of 2,4-dinitrophenol on the differentiation of the slime mold *Dictyostelium discoideum. J. Cellular Comp. Physiol. 37*: 323-336.

Hirschy, R. A. and K. B. Raper (1964). Light control of macrocyst formation in *Dictyostelium* (Abstract). *Bacteriol. Proc.* p. 27.

Hohl, H. R. (1965). Nature and development of membrane systems in food vacuoles of cellular slime molds predatory upon bacteria. *J. Bacteriol. 90*: 755-765.

Hohl, H. R. and K. B. Raper (1963a). Nutrition of cellular slime molds. I. Growth on living and dead bacteria. *J. Bacteriol. 85*: 191-198.

Hohl, H. R. and K. B. Raper (1963b). Nutrition of cellular slime molds. II. Growth of *Polysphondylium pallidum* in axenic culture. *J. Bacteriol. 85*: 199-206.

Hohl, H. R. and K. B. Raper (1963c). Nutrition of cellular slime molds. III. Specific growth requirements of *Polysphondylium pallidum. J. Bacteriol. 86*: 1314-1320.

Hohl, H. R. and K. B. Raper (1964). Control of sorocarp size in the cellular slime mold *Dictyostelium discoideum. Develop. Biol. 9*: 137-153.

Hostak, M. B. (1960). Induced aggregation of myxamoebae in *Acytostelium*. M.S. thesis, University of Wisconsin.

Hostak, M. B. and K. B. Raper (1960). The induction of cell aggregation in *Acytostelium* by alkaloids (Abstract). *Bacteriol. Proc.* pp. 58-59.

Huffman, D. M., A. J. Kahn, and L. S. Olive (1962). Anastomosis and cell fusions in *Dictyostelium*. *Proc. Natl. Acad. Sci. U.S. 48*: 1160-1164.

Huffman, D. M. and L. S. Olive (1963). A significant morphogenetic variant of *Dictyostelium mucoroides*. *Mycologia 55*: 333-344.

Huffman, D. M. and L. S. Olive (1964). Engulfment and anastomosis in the cellular slime molds (Acrasiales). *Am. J. Botany 51*: 465-471.

Jaffe, L. F. (1958). Morphogenesis in lower plants. *Ann. Rev. Plant Physiol. 9*: 359-384.

Johnson, D. F., B. E. Wright, and E. Heftmann (1962). Biogenesis of delta[22]-stigmasten-3 beta-ol in *Dictyostelium discoideum*. *Arch. Biochem. Biophys. 97*: 232-235.

Kabiersch, W. (1941). Die Entwicklung von *Dictyostelium mucoroides*. (Commentary on the film of A. Arndt c-381/1941.) Reichsanstalt für Film und Bild in Wissenschaft und Unterricht Hochschulfilm.

Kahn, A. J. (1964a). Some aspects of cell interaction in the development of the slime mold *Dictyostelium purpureum*. *Develop. Biol. 9*: 1-19.

Kahn, A. J. (1964b). The influence of light on cell aggregation in *Polysphondylium pallidum*. *Biol. Bull. 127*: 85-96.

Kessler, D. and K. B. Raper (1960). *Guttulina*, a rediscovered genus of cellular slime mold. (Abstract). *Bacteriol. Proc.*, p. 58.

Kitzke, E. D. (1948). Two members of the Acrasieae isolated in Milwaukee County, Wisconsin. *Papers Mich. Acad. Sci. 34*: 13-18.

Kitzke, E. D. (1949). Some ecological aspects of the Acrasiales in and near Madison, Wisconsin. *Papers Mich. Acad. Sci. 35*: 25-32.

Kitzke, E. D. (1952). A new method for isolating members of the Acrasieae from soil samples. *Nature 170*: 284-285.

Konijn, T. M. (1966a). Chemotaxis in the cellular slime molds. I. The effect of temperature. *Develop. Biol. 12*: 487-497.

Konijn, T. M. (1966b). Chemotaxis in the cellular slime molds. II. The effect of density. (In press.)

Konijn, T. M. and K. B. Raper (1961). Cell aggregation in *Dictyostelium discoideum. Develop. Biol. 3*: 725-756.

Konijn, T. M. and K. B. Raper (1965). The influence of light on the time of cell aggregation in the *Dictyosteliaceae. Biol. Bull. 128*: 392-400.

Kostellow, A. (1956). Developmental response of *Dictyostelium discoideum* to some amino acids and their analogues. Ph.D. thesis, Columbia University.

Krichevsky, M. I. and L. L. Love (1964a). The uptake and utilization of histidine by washed amoebae in the course of development in *Dictyostelium discoideum. J. Gen. Microbiol. 34*: 483-490.

Krichevsky, M. I. and L. L. Love (1964b). Adenine inhibition of the rate of sorocarp formation in *Dictyostelium discoideum. J. Gen. Microbiol. 37*: 293-295.

Krichevsky, M. I. and L. L. Love (1965). Efflux of macromolecules from washed *Dictyostelium discoideum. J. Gen. Microbiol. 41*: 367-374.

Krichevsky, M. I. and B. E. Wright (1963). Environmental control of the course of development in *Dictyostelium discoideum. J. Gen. Microbiol. 32*: 195-207.

Krivanek, J. O. (1956). Alkaline phosphatase activity in the developing slime mold, *Dictyostelium discoideum* Raper. *J. Exptl. Zool. 133*: 459-480.

Krivanek, J. O. (1964). Nucleic acids in the developing slime mold, *Dictyostelium discoideum* (Abstract). *Bull. Assoc. Southeastern Biol. 11*: 49.

Krivanek, J. O. and R. C. Krivanek (1957). A method for embedding small specimens. *Stain Technol. 32*: 300-301.

Krivanek, J. O. and R. C. Krivanek (1958). The histochemical localization of certain biochemical intermediates and enzymes in the developing slime mold, *Dictyostelium discoideum* Raper. *J. Exptl. Zool. 137*: 89-115.

Krivanek, J. O. and R. C. Krivanek (1959). Chromatographic analyses of amino acids in the developing slime mold, *Dictyostelium discoideum* Raper. *Biol. Bull. 116*: 265-271.

Krivanek, J. O. and R. C. Krivanek (1962). Evidence for the occurrence of transamination in the developing slime mold,

Dictyostelium discoideum (Abstract). Am. Zoologist 2: 421.

Krivanek, J. O. and R. C. Krivanek (1965). Evidence for transaminase activity in the slime mold, *Dictyostelium discoideum* Raper. *Biol. Bull. 129*: 295-302.

Krzemieniewski, H. S. (1927). Z. Mikroflory gleby w Polace. (Contributions à la microflore du sol en Pologne.) *Acta Soc. Botan. Polon. 4*: 141-144.

Labudde, B. F. (1956). A cytological study of *Dictyostelium*. Master's thesis, University of Wisconsin.

Lang, A. (1954). Entwicklungsphysiologie der Acrasiales. *Fortschr. Botan. 15*: 400-475.

Liddel, G. U. and B. E. Wright (1961). The effect of glucose on respiration of the differentiating slime mold. *Develop. Biol. 3*: 265-276.

Lonert, A. C. (1965). A week-end with a cellular slime mold. *Turtox News 43*: 50-53.

Marchal, E. (1885). Champignons coprophiles de la Belgique. *Bull. Soc. Roy. Botan. Belg. 24*: 57-74 (see p. 74 and Pl. 3, Figs. 1-4).

Mercer, E. H. and B. M. Shaffer (1960). Electron microscopy of solitary and aggregated slime mould cells. *J. Biophys. Biochem. Cytol. 7*: 353-356.

Michalska, I. and F. X. Skupienski (1939). Recherches écologique sur les Acrasiées *Polysphondylium pallidum* Olive, *Polysphondylium violaceum* Bref., *Dictyostelium mucoroides* Bref. *Compt. Rend. Acad. Sci. 207*: 1239-1241.

Morse, M. (1949). Equilibria in nature—stable and unstable. *Proc. Am. Phil. Soc. 93*: 222-225.

Mühlethaler, K. (1956). Electron microscopic study of the slime mold *Dictyostelium discoideum. Am. J. Botany 43*: 673-678.

Nadson, G. A. (1899-1900). Des cultures du *Dictyostelium mucoroides* Bref. et des cultures pures des amibes en général. *Scripta Botan. Hort. Univ. Imp. Petropolitanae 15*: 188-190. (In Russian. Résumé in French.)

Oehler, R. (1922). *Dictyostelium mucoroides* (Brefeld). *Zentr. Bakteriol. Parasitenk. 89*: 155-156.

Olive, E. W. (1901). Preliminary enumeration of the Sorophoreae. *Proc. Am. Acad. Sci. 37*: 333-344.

Olive, E. W. (1902). Monograph of the Acrasieae. *Proc. Boston Soc. Nat. Hist. 30*: 451-513.

Olive, L. S. (1962). The genus *Protostelium. Am. J. Botany 49*: 297-303.

Olive, L. S. (1963). The question of sexuality in cellular slime molds. *Bull. Torrey Botan. Club 90*: 144-147.

Olive, L. S. (1964). A new member of the Mycetozoa. *Mycologia 61*: 885-896.

Olive, L. S. (1965). A developmental study of *Guttulinopsis vulgaris* (Acrasiales). *Am. J. Botany 52*: 513-519.

Olive, L. S., S. K. Dutta, and C. Stoianovitch (1961). Variation in the cellular slime mold *Acrasis rosea. J. Protozool. 8*: 467-472.

Olive, L. S. and C. Stoianovitch (1960). Two new members of the Acrasiales. *Bull. Torrey Botan. Club 87*: 1-20.

Olive, L. S. and C. Stoianovitch (1966a). A new mycetozoan genus intermediate between *Cavostelium* and *Protostelium*; a new order of Mycetozoa. *J. Protozool. 13*: 164-171.

Olive, L. S. and C. Stoianovitch (1966b) A simple new mycetozoan with ballistospores. *Am. J. Botany 53*: 344-349.

Olive, L. S. and C. Stoianovitch (1966c). A two-spored species of *Cavostelium* (Protostelida). *Mycologia 58*: 440-451.

Olive, L. S. and C. Stoianovitch (1966d). *Protosteliopsis,* a new genus of Protostelida. *Mycologia 58*: 452-455.

Opderbeck, C. T. (1961). The effect of histidine on aggregation in *Dictyostelium purpureum*. Senior thesis, Princeton University.

Oudemans, C. A. J. A. (1885). Aanwinstein voor de flora mycologica van Nederland. *Ned. botan. ver.,* Leyden. *Ned. Kruidkundig Arch; Verslagen en Mededeel.* 2nd series *4*: 241-242 (Pl. 4, Fig. 4).

Paddock, R. B. (1953). The appearance of amoebae tracks in cultures of *Dictyostelium discoideum. Science 118*: 597-598.

Palm, B. T. (1935). Ett fynd av *Dictyostelium mucoroides* i Sydsverige (*D. mucoroides* from South Sweden). *Svenska Botan. Tidskr. 29*: 365-366. (English summary.)

Pavillard, J. (1953). Ordre des Acrasiés. Pp. 493-505 in *Traité de Zoologie,* Vol. I/II. P. Grassé, ed. Masson & Co., Paris.

Pfützner-Eckert, R. (1950). Entwicklungsphysiologische Untersuchungen an *Dictyostelium mucoroides* Brefeld. *Arch. Entwicklungsmech. Organ. 144*: 381-409.

Phillips, W. D., A. Rich, and R. R. Sussman (1964). The isolation and identification of polyribosomes from cellular slime molds. *Biochim. Biophys. Acta 80*: 508-510.

Pinoy, E. (1903). Nécessité d'une symbiose microbienne pour obtenir la culture des myxomycètes. *Compt. Rend. Acad. Sci. 137*: 580-581.

Pinoy, E. (1907). Rôle des bactéries dans le développment de certains myxomycètes. *Ann. Inst. Pasteur 21*: 622-656; 686-700.

Pinoy, P. E. (1950). Quelques observations sur la culture d'une Acrasiée. *Bull. Soc. Mycol. France 66*: 37-38.

Potts, G. (1902). Zur Physiologie des *Dictyostelium mucoroides*. *Flora 91*: 281-347.

Rafaeli, D. C. (1962). Studies on mixed morphological mutants of *Polysphondylium violaceum*. *Bull. Torrey Botan. Club 89*: 312-318.

Rai, J. N. and J. P. Tewari (1961). Studies in cellular slime moulds from Indian soils. I. On the occurrence of *Dictyostelium mucoroides* Bref. and *Polysphondylium violaceum* Bref. *Proc. Indian Acad. Sci. 53*: 1-9.

Rai, J. N. and J. P. Tewari (1963a). Studies in cellular slime moulds from Indian soils. II. On the occurrence of an aberrant strain of *Polysphondylium violaceum* Bref., with a discussion on the relevance of mode of branching of the sorocarp as a criterion for classifying members of Dictyosteliaceae. *Proc. Indian Acad. Sci. 58*: 201-206.

Rai, J. N. and J. P. Tewari (1963b). Studies in cellular slime moulds from Indian soils. III. On the occurrence of two strains of *Dictyostelium mucoroides*-complex, conforming to the species *Dictyostelium sphaerocephalum* (Oud.) Saccardo and March. *Proc. Indian Acad. Sci. 58*: 263-266.

Raper, K. B. (1935). *Dictyostelium discoideum*, a new species of slime mold from decaying forest leaves. *J. Agr. Res. 50*: 135-147.

Raper, K. B. (1937). Growth and development of *Dictyostelium discoideum* with different bacterial associates. *J. Agr. Res. 55*: 289-316.

Raper, K. B. (1939). Influence of culture conditions upon the growth and development of *Dictyostelium discoideum*. *J. Agr. Res. 58*: 157-198.

Raper, K. B. (1940a). The communal nature of the fruiting process in the Acrasieae. *Am. J. Botan. 27*: 436-448.

Raper, K. B. (1940b). Pseudoplasmodium formation and organization in *Dictyostelium discoideum. J. Elisha Mitchell Sci. Soc. 56*: 241-282.

Raper, K. B. (1941a). *Dictyostelium minutum*, a second new species of slime mold from decaying forest leaves. *Mycologia 33*: 633-649.

Raper, K. B. (1941b). Developmental patterns in simple slime molds. Third Growth Symposium. *Growth 5*: 41-76.

Raper, K. B. (1951). Isolation, cultivation, and conservation of simple slime molds. *Quart. Rev. Biol. 26*: 169-190.

Raper, K. B. (1956a). Factors affecting growth and differentiation in simple slime molds. *Mycologia 48*: 169-205.

Raper, K. B. (1956b). *Dictyostelium polycephalum* n. sp.: a new cellular slime mould with coremiform fructifications. *J. Gen. Microbiol. 14*: 716-732.

Raper, K. B. (1960a). Levels of cellular interaction in amoeboid populations. *Proc. Am. Phil. Soc. 104*: 579-604.

Raper, K. B. (1960b). Acrasiales. *McGraw-Hill Encyclopedia of Science and Technology 1*: 49-50.

Raper, K. B. (1963). The environment and morphogenesis in cellular slime molds. *Harvey Lectures. 57*: 111-141. Academic Press, Inc., New York.

Raper, K. B. and D. I. Fennell (1952). Stalk formation in *Dictyostelium. Bull. Torrey Botan. Club 79*: 25-51.

Raper, K. B. and M. S. Quinlan (1958). *Acytostelium leptosomum*: a unique cellular slime mold with an acellular stalk. *J. Gen. Microbiol. 18*: 16-32.

Raper, K. B. and N. R. Smith (1939). The growth of *Dictyostelium discoideum* upon pathogenic bacteria. *J. Bacteriol. 38*: 431-444.

Raper, K. B. and C. Thom (1932). The distribution of *Dictyostelium* and other slime molds in soil. *J. Wash. Acad. Sci. 22*: 93-96.

Raper, K. B. and C. Thom (1941). Interspecific mixtures in the Dictyosteliaceae. *Am. J. Botany 28*: 69-78.

Ray, D. L. and R. E. Hayes (1954). *Hartmanella astronyxis*: a new species of free-living amoeba. *J. Morphol. 95*: 159-188.

Reinhardt, D. J. (1966). The social amoebae (cellular slime molds). *Turtox News 44*: 50-56.

Rorke, J. and G. Rosenthal (1959). Influences on the spatial arrangements of *Dictyostelium discoideum.* Senior thesis, Princeton University.

Rosen, O. M., S. M. Rosen, and B. L. Horecker (1965). Fate of the cell wall of *Salmonella typhimurium* upon ingestion by the cellular slime mold, *Polysphondylium pallidum. Biochem. Biophys. Res. Commun. 18*: 270-276.

Ross, I. K. (1960). Studies on diploid strains of *Dictyostelium discoideum Am. J. Botany 47*: 54-59.

Runyon, E. H. (1942). Aggregation of separate cells of *Dictyostelium* to form a multicellular body. *Collecting Net 17*: 88.

Russell, G. K. and J. T. Bonner (1960). A note on spore germination in the cellular slime mold *Dictyostelium mucoroides. Bull. Torrey Botan. Club 87*: 187-191.

Samuel, E. W. (1961). Orientation and rate of locomotion of individual amebas in the life cycle of the cellular slime mold *Dictyostelium mucoroides. Develop. Biol. 3*: 317-335.

Schildkraut, C. L., M. Mandel, S. Levisohn, J. E. Smith-Sonneborn, and J. Marmur (1962). Deoxyribonucleic acid base composition and taxonomy of some protozoa. *Nature 196*: 795-796.

Schuckmann, W. von (1924). Zur Biologie von *Dictyostelium mucoroides* Bref. *Zent. Bakteriol. Parasitenk. 91*: 302-309.

Schuckmann, W. von (1925). Zur Morphologie und Biologie von *Dictyostelium mucoroides* Bref. *Arch. Protistenk. 51*: 495-529.

Shaffer, B. M. (1953). Aggregation in cellular slime moulds: *in vitro* isolation of acrasin. *Nature 171*: 975.

Shaffer, B. M. (1956a). Properties of acrasin. *Science 123*: 1172-1173.

Shaffer, B. M. (1956b). Acrasin, the chemotactic agent in cellular slime moulds. *J. Exptl. Biol. 33*: 645-657.

Shaffer, B. M. (1957a). Aspects of aggregation in cellular slime moulds. I. Orientation and chemotaxis. *Am. Naturalist 91*: 19-35.

Shaffer, B. M. (1957b). Properties of slime-mould amoebae of significance for aggregation. *Quart. J. Microscop. Sci. 98*: 377-392.

Shaffer, B. M. (1957c). Variability of behaviour of aggregating cellular slime moulds. *Quart. J. Microscop. Sci. 98*: 393-405.

Shaffer, B. M. (1958). Integration in aggregating cellular slime moulds. *Quart. J. Microscop. Sci.* *99*: 103-121.

Shaffer, B. M. (1961a). Species differences in the aggregation of the Acrasieae. Pp. 294-298 in *Recent Advances in Botany*, Proc. 9th Intern. Botan. Congr. University of Toronto Press.

Shaffer, B. M. (1961b). The cells founding aggregation centres in the slime mould *Polysphondylium violaceum. J. Exptl. Biol.* *38*: 833-849.

Shaffer, B. M. (1962) & (1964c). The Acrasina. *Advan. Morphogenesis 2*: 109-182; *3*: 301-322.

Shaffer, B. M. (1963a). Inhibition by existing aggregations of founder differentiation in the cellular slime mould *Polysphondylium violaceum. Exptl. Cell Res. 31*: 432-435.

Shaffer, B. M. (1963b). Behaviour of particles adhering to amoebae of the slime mould *Polysphondylium violaceum* and the fate of the cell surface during locomotion. *Exptl. Cell Res. 32*: 603-606.

Shaffer, B. M. (1964a). Intracellular movement and locomotion of cellular slime-mold amebae. Pp. 387-405 in *Primitive Motile Systems in Cell Biology*, R. D. Allen and N. Kamiya, eds. Academic Press, Inc., New York.

Shaffer, B. M. (1964b). Attraction through air exerted by unaggregated cells on aggregates of the slime mould *Polysphondylium violaceum. J. Gen. Microbiol. 36*: 359-364.

Shaffer, B. M. (1964c). See Shaffer (1962).

Shaffer, B. M. (1965a). Antistrophic pseudopodia of the collective amoeba *Polysphondylium violaceum. Exptl. Cell Res. 37*: 79-92.

Shaffer, B. M. (1965b). Mechanical control of the manufacture and resorption of cell surface in collective amoebae. *J. Theoret. Biol. 8*: 27-40.

Shaffer, B. M. (1965c). Cell movement within aggregates of the slime mould *Dictyostelium discoideum* revealed by surface markers. *J. Embryol. Exptl. Morphol. 13*: 97-117.

Shaffer, B. M. (1965d). Pseudopodia and intracytoplasmic displacements of the collective amoebae Dictyosteliidae. *Exptl. Cell Res. 37*: 12-25.

Shoopman, J. (1963). *Dictyostelium diminutivum*, sp. Nov. A new cellular slime mold from Mexican soils. M.S. thesis, University of Wisconsin.

Singh, B. N. (1946). Soil Acrasieae and their bacterial food supply. *Nature 157*: 133-134.

Singh, B. N. (1947a). Studies on soil Acrasieae: 1. Distribution of species of *Dictyostelium* in soils of Great Britain and the effect of bacteria on their development. *J. Gen. Microbiol. 1*: 11-21.

Singh, B. N. (1947b). Studies on soil Acrasieae: 2. The active life of species of *Dictyostelium* in soil and the influence thereon of soil moisture and bacterial food. *J. Gen. Microbiol. 1*: 361-367.

Skupienski, F. X. (1919). Sur la sexualité chez une espèce de Myxomycète Acrasiée, *Dictyostelium mucoroides. Compt. Rend. Acad. Sci. 167*: 960-962.

Skupienski, F. X. (1920). *Recherches sur le Cycle Évolutif de Certains Myxomycètes*. Paris. 81 pp.

Slifkin, M. K. and J. T. Bonner (1952). The effect of salts and organic solutes on the migration time of the slime mold *Dictyostelium discoideum. Biol. Bull. 102*: 273-277.

Slifkin, M. K. and H. S. Gutowsky (1958). Infrared spectroscopy as a new method for assessing the nutritional requirements of the slime mold, *Dictyostelium discoideum. J. Cellular Comp. Physiol. 51*: 249-257.

Snyder, H. M. and C. Ceccarini (1966). Interspecific spore inhibition in the cellular slime molds. *Nature. 209*: 1152.

Solomon, E. P., E. M. Johnson, and J. H. Gregg (1964). Multiple forms of enzymes in a cellular slime mold during morphogenesis. *Develop. Biol. 9*: 314-326.

Sonneborn, D. R., L. Levine, and M. Sussman (1965). Serological analyses of cellular slime mold development. II. Preferential loss, during morphogenesis, of antigenic activity associated with the vegetative myxamoebae. *J. Bacteriol. 89*: 1092-1096.

Sonneborn, D. R., M. Sussman, and L. Levine (1964). Serological analyses of cellular slime-mold development. I. Changes in antigenic activity during cell aggregation. *J. Bacteriol. 87*: 1321-1329.

Sonneborn, D. R., G. J. White, and M. Sussman (1963). A mutation affecting both rate and pattern of morphogenesis in *Dictyostelium discoideum. Develop. Biol. 7*: 79-93.

Staples, S. (1965). The formation and properties of a pigment in the cellular slime molds. Senior thesis, University of Florida.

Staples, S. O. and J. H. Gregg (1966). Carotenoid pigments in the cellular slime mold, *Dictyostelium discoideum*. *Biol. Bull.* (In press.)

Stong, C. L. (1966). How to cultivate the slime molds and perform experiments on them. *Sci. Am. 214*: 116-121.

Sussman, M. (1951). The origin of cellular heterogeneity in the slime molds, Dictyosteliaceae. *J. Exptl. Zool. 118*: 407-418.

Sussman, M. (1952). An analysis of the aggregation stage in the development of the slime molds, Dictyosteliaceae. II. Aggregative center formation by mixtures of *Dictyostelium discoideum* wild type and aggregateless variants. *Biol. Bull. 103*: 446-457.

Sussman, M. (1954). Synergistic and antagonistic interactions between morphogenetically deficient variants of the slime mould *Dictyostelium discoideum. J. Gen. Microbiol. 10*: 110-120.

Sussman, M. (1955a). "Fruity" and other mutants of the cellular slime mould, *Dictyostelium discoideum*: a study of developmental aberrations. *J. Gen. Microbiol. 13*: 295-309.

Sussman, M. (1955b). The developmental physiology of the amoeboid slime molds. Pp. 201-223 in *Biochemistry and Physiology of the Protozoa*, Vol. 2, S. Hutner and A. Lwoff, eds., Academic Press, New York.

Sussman, M. (1956a). On the relation between growth and morphogenesis in the slime mold *Dictyostelium discoideum. Biol. Bull. 110*: 91-95.

Sussman, M. (1956b). The biology of the cellular slime molds. *Ann. Rev. Microbiol. 10*: 21-50.

Sussman, M. (1958). A developmental analysis of cellular slime mold aggregation. Pp. 264-295 in *A Symposium on the Chemical Basis of Development*, W. D. McElroy and B. Glass, eds. Johns Hopkins Press.

Sussman, M. (1961a). Cultivation and serial transfer of the slime mould, *Dictyostelium discoideum* in liquid nutrient medium. *J. Gen. Microbiol. 25*: 375-378.

Sussman, M. (1961b). Cellular differentiation in the slime mold. Pp. 221-239 in *Growth in Living Systems*, M. X. Zarrow, ed. Basic Books, New York.

Sussman, M. (1963). Growth of the cellular slime mold *Poly-sphondylium pallidum* in a simple nutrient medium. *Science 139*: 338.

Sussman, M. (1965a). Inhibition by actidione of protein synthesis and UDP-Gal polysaccharide transferase accumulation in *Dictyostelium discoideum*. *Biochem. Biophys. Res. Commun. 18*: 763-767.

Sussman, M. (1965b). Temporal, spatial, and quantitative control of enzyme activity during slime mold cytodifferentiation. Pp. 66-76 in *Genetic Control of Differentiation*, Brookhaven Symposia in Biology No. 18.

Sussman, M. (1966). Protein synthesis and the temporal control of genetic transcription during slime mold development. *Proc. Natl. Acad. Sci. U.S. 55*: 813-818.

Sussman, M. and S. G. Bradley (1954). A protein growth factor of bacterial origin required by the cellular slime molds. *Arch. Biochem. Biophys. 51*: 428-435.

Sussman, M. and H. L. Ennis (1959). The role of the initiator cell in slime mold aggregation. *Biol. Bull. 116*: 304-317.

Sussman, M. and F. Lee (1954). Physiology of developmental variants among the cellular slime molds (Abstract). *Bacteriol. Proc.*, p. 42.

Sussman, M. and F. Lee (1955). Interactions among variant and wild-type strains of cellular slime molds across thin agar membranes. *Proc. Natl. Acad. Sci. U.S. 41*: 70-78.

Sussman, M., F. Lee, and N. S. Kerr (1956). Fractionation of acrasin, a specific chemotactic agent for slime mold aggregation. *Science 123*: 1171-1172.

Sussman, M. and N. Lovgren (1965). Preferential release of the enzyme UDP-galactose polysaccharide transferase during cellular differentiation in the slime mold, *Dictyostelium discoideum*. *Exptl. Cell Res. 38*: 97-105.

Sussman, M. and E. Noël (1952). An analysis of the aggregation stage in the development of the slime molds, Dictyosteliaceae. I. The populational distribution of the capacity to initiate aggregation. *Biol. Bull. 103*: 259-268.

Sussman, M. and M. J. Osborn (1964). UDP-galactose polysaccharide transferase in the cellular slime mold, *Dictyostelium discoideum*: appearance and disappearance of activity during cell differentiation. *Proc. Natl. Acad. Sci. U.S. 52*: 81-87.

Sussman, M. and R. R. Sussman (1956). Cellular interactions during the development of the cellular slime molds. Pp. 125-154 in *Cellular Mechanisms in Differentiation and Growth*, 14th Growth Symposium, D. Rudnick, ed. Princeton University Press.

Sussman, M. and R. R. Sussman (1961). Aggregative performance. *Exptl. Cell Res. Suppl. 8*: 91-106.

Sussman, M. and R. R. Sussman (1962). Ploidal inheritance in *Dictyostelium discoideum*. I: Stable haploid, stable diploid and metastable strains. *J. Gen. Microbiol. 28*: 417-429.

Sussman, M. and R. R. Sussman (1965). The regulatory program for UDP-Gal polysaccharide transferase activity during slime mold cytodifferentiation: Requirement for specific synthesis of RNA. *Biochim. Biophys. Acta 108*: 463-473.

Sussman, R. R. (1961). A method for staining chromosomes of *D. discoideum* myxamoebae in the vegetative stage. *Exptl. Cell Res. 24*: 154-155.

Sussman, R. R. and M. Sussman (1953). Cellular differentiation in Dictyosteliaceae: heritable modifications of the developmental pattern. *Ann. N. Y. Acad. Sci. 56*: 949-960.

Sussman, R. R. and M. Sussman (1960). The dissociation of morphogenesis from cell division in the cellular slime mould, *Dictyostelium discoideum*. *J. Gen. Microbiol. 23*: 287-293.

Sussman, R. R. and M. Sussman (1963). Ploidal inheritance in the slime mould *Dictyostelium discoideum*: haploidization and genetic segregation of diploid strains. *J. Gen. Microbiol. 30*: 349-355.

Sussman, R. R., M. Sussman, and H. L. Ennis (1960). Appearance and inheritance of the I-cell phenotype in *D. discoideum*. *Develop. Biol. 2*: 367-392.

Sussman, R. R., M. Sussman, and F. L. Fu (1958). The chemotactic complex responsible for cellular slime mold aggregation (Abstract). *Bacteriol. Proc.*, p. 32.

Takeuchi, I. (1960). The correlation of cellular changes with succinic dehydrogenase and cytochrome oxidase activities in the development of the cellular slime molds. *Develop. Biol. 2*: 343-366.

Takeuchi, I. (1963). Immunochemical and immunohistochemical studies on the development of the cellular slime mold *Dictyostelium mucoroides*. *Develop. Biol. 8*: 1-26.

Takeuchi, I. and M. Tazawa (1955). Studies on the morphogenesis of the slime mould, *Dictyostelium discoideum*. *Cytologia 20*: 157-165.

Thom, C. and K. B. Raper (1930). Myxamoebae in soil and decomposing crop residues. *J. Wash. Acad. Sci. 20*: 362-370.

Tieghem, P. van (1880). Sur quelques Myxomycètes à plasmode agrégé. *Bull. Soc. Botan. France 27*: 317-322.

Tieghem, P. van (1884). *Coenonia,* genre nouveau de Myxomycètes à plasmode agrégé. *Bull. Soc. Botan. France 31*: 303-306.

Vuillemin, P. (1903). Une Acrasiée bactériophage. *Compt. Rend. Acad. Sci. 137*: 387-389.

Ward, C. and B. E. Wright (1965). Cell wall synthesis in *Dictyostelium discoideum*. I. In vitro synthesis from uridine diphosphoglucose. *Biochemistry 4*: 2021-2027.

Ward, J. M. (1958). Biochemical systems involved in differentiation of the fungi. Pp. 33-64 in *Fourth International Congress of Biochemistry*, Vol. VI, *Biochemistry of Morphogenesis*. Pergamon Press.

Weinkauff, A. M. and M. F. Filosa (1965). Factors involved in the formation of macrocysts by the cellular slime mold, *Dictyostelium mucoroides*. *Can. J. Microbiol. 11*: 385-387.

Weitzman, I. (1962). Studies on the nutrition of *Acrasis rosea*. *Mycologia 54*: 113-115.

Wescott, B. A. (1960). The effect of changes in the gaseous environment upon the growth and development of *Dictyostelium discoideum*. M.S. thesis, University of Wisconsin.

White, G. J. and M. Sussman (1961). Metabolism of major cell constituents during slime mold morphogenesis. *Biochim. Biophys. Acta 53*: 285-293.

White, G. J. and M. Sussman (1963a). Polysaccharides involved in slime mold development. I. Water-soluble glucose polymer(s). *Biochim. Biophys. Acta 74*: 173-178.

White, G. J. and M. Sussman (1963b). Polysaccharides involved in slime mold development. II. Water-soluble acid mucopolysaccharide(s). *Biochim. Biophys. Acta 74*: 179-187.

Whitfield, F. E. (1964). The use of proteolytic and other enzymes in the separation of slime mould grex. *Exptl. Cell Res. 36*: 62-72.

Whittingham, W. F. and K. B. Raper (1956). Inhibition of normal pigment synthesis in spores of *Dictyostelium purpureum*. *Am. J. Botany 43*: 703-708.

Whittingham, W. F. and K. B. Raper (1957). Environmental factors influencing the growth and fructification of *Dictyostelium polycephalum*. *Am. J. Botany 44*: 619-627.

Whittingham, W. F. and K. B. Raper (1960). Non-viability of stalk cells in *Dictyostelium*. *Proc. Natl. Acad. Sci. U.S. 46*: 642-649.

Wilson, C. M. (1952). Sexuality in the Acrasiales. *Proc. Natl. Acad. Sci. U.S. 38*: 659-662.

Wilson, C. M. (1953). Cytological study of the life cycle of *Dictyostelium*. *Am. J. Botany 40*: 714-718.

Wilson, C. M. and I. K. Ross (1957). Further cytological studies in the Acrasiales. *Am. J. Botany 44*: 345-350.

Wright, B. E. (1958). Effect of steroids on aggregation in the slime mold *Dictyostelium discoideum* (Abstract). *Bacteriol. Proc.* p. 115.

Wright, B. E. (1960). On enzyme-substrate relationships during biochemical differentiation. *Proc. Natl. Acad. Sci. U.S. 46*: 798-803.

Wright, B. (1963a). Symposium on biochemical bases of morphogenesis in fungi. I. Endogenous substrate control in biochemical differentiation. *Bacteriol. Rev. 27*: 273-281.

Wright, B. E. (1963b). Endogenous activity and sporulation in slime molds. *Ann. N. Y. Acad. Sci. 102*: 740-754.

Wright, B. E. (1964). Biochemistry of Acrasiales. Pp. 341-381 in *Biochemistry and Physiology of the Protozoa*, Vol. 3, S. H. Hutner, ed. Academic Press, Inc., New York.

Wright, B. E. (1965). Control of carbohydrate synthesis in the slime mold. Pp. 298-316 in *Development and Metabolic Control and Neoplasia*. The Williams and Wilkins Company, Baltimore.

Wright, B. E. and M. L. Anderson (1958). Enzyme patterns during differentiation in the slime mold. Pp. 296-314 in *A Symposium on the Chemical Basis of Development*, W. D. McElroy and B. Glass, eds. Johns Hopkins Press.

Wright, B. E. and M. L. Anderson (1959). Biochemical differentiation in the slime mold. *Biochim. Biophys. Acta 31*: 310-322.

Wright, B. E. and M. L. Anderson (1960a). Protein and amino acid turnover during differentiation in the slime mold. I.

Utilization of endogenous amino acids and proteins. *Biochim. Biophys. Acta 43*: 62-66.

Wright, B. E. and M. L. Anderson (1960b). Protein and amino acid turnover during differentiation in the slime mold. II. Incorporation of 35 S methionine into the amino acid pool and into protein. *Biochim. Biophys. Acta 43*: 67-78.

Wright, B. E. and S. Bard (1963). Glutamate oxidation in the differentiating slime mold. I. Studies *in vivo. Biochim. Biophys. Acta 71*: 45-49.

Wright, B. E. and B. Bloom (1960). *In vivo* investigations of glucose catabolism in a differentiating slime mold (Abstract). *Bacteriol. Proc.*, p. 59.

Wright, B. E. and B. Bloom (1961). *In vivo* evidence for metabolic shifts in the differentiating slime mold. *Biochim. Biophys. Acta 48*: 342-346.

Wright, B. E. and M. Brühmüller (1964). The effect of exogenous glucose concentration of C-6/C-1 ratio. *Biochim. Biophys. Acta 82*: 203-204.

Wright, B. E., M. Brühmüller, and C. Ward (1964). Studies *in vivo* on hexose metabolism in *Dictyostelium discoideum. Develop. Biol. 9*: 287-297.

Wright, B. E. and C. Ward (1966). Cell wall synthesis in *Dictyostelium discoideum*. I. *In vitro* synthesis from uridine diphosphoglucose. *Biochemistry*. (In press.)

Wright, B. E., C. Ward, and D. Dahlberg (1966). Cell wall polysaccharide synthesis *in vitro* catalyzed by enzymes from myxamoebae of *Dictyostelium discoideum. Biochem. Biophys. Res. Commun. 22*: 352-356.

Wright, B. E. and M. E. Wassarman (1964). Pyridine nucleotide levels in *Dictyostelium discoideum* during differentiation. *Biochim. Biophys. Acta 90*: 423-424.

Zaczynski, E. J. (1951). The effect of anti-serum on the growth and morphogenesis of the slime mold *Dictyostelium discoideum*. M.S. thesis. Vanderbilt University.

Zopf, W. (1885). Die Pilzthiere oder Schleimpilze. Reprinted from *Die Encyklopaedie der Naturwissenschaften*, Breslau, pp. 1-174.

Index

INDEX

Hartmanella, 20, 21, *22*, 24, 74;
  amoeba of, 20, 21
Hartmann, M., 25
Hayes, R. E., 20, 21, 24
heat, *see* temperature
heterocaryosis, 16, 25
heterocytosis, 17, 168
Hirschy, R. A., 138
histidine, effect on aggregation, 140
Hoffman, M. E., 51, 52, 54, 136,
  141, 143
Hohl, H. R., 87, 88, 148ff
Hollande, A., 5
Holtfreter, J., 161
Huffman, D. M., 11, 54
humidity, effect of, 69, 78; effect
  on aggregation, 133; effect on migration, 134; effect on spore differentiation, 135; requirements
  of *D. polycephalum*, 69, 137
Huxley, J. S., 160
hydrotaxis, 120

immunological techniques, 131, 155,
  173
inhibitor substance, 52, 136
initiator cells, 139, 156
interphase, between vegetative and
  aggregation stages, 50, 52, 129

Jennings, H. S., 170

Kahn, A. J., 11, 54, 133, 143, 144,
  166
Karling, J. S., 7
Kelso, A. P., 95, 114
Kerr, N. S., 109, 113
Kolderie, M. Q., 45, 48, 55, 58, 60,
  155, 170
Konijn, T. M., 124, 134, 139
Konijn's test, for acrasin, 124
Koontz, P. G., 95
Kostellow, A., 164
Krichevsky, M. I., 140
Krivanek, J. O., 129
Kühn, A., 107

Labudde, B. F., 45, 53
Labyrinthula, 4, 5, *6*

Lederberg, J., 171
Ledingham, G. A., 7
Lee, F., 109, 113, 166
Levine, L., 131
light, effect on aggregation, 71, 133,
  144; orientation to, 118
Lonert, A. C., 82
Love, L. L., 140
Lovgren, N., 130

macrocysts, 40ff, *42*, 138, 159, 167
McVittie, A., 15
media, for growth, 81ff
meiosis, 53ff
Mercer, E. H., 45
metachromatic granules, 46, 49
microcysts, 28, 30, 40, *42*
migrating pseudoplasmodia, *102*,
  plate 5; orientation of, 118, *119*
migration, of *D. polycephalum*, 67;
  duration of, 134
mitochondria, 50, 129
mitosis, 53
morphogenetic movement, 76, 92ff
Moscona, A., 161
Mühlethaler, K., 45, 46, 57, 66
multinucleate cells, 54
mutants, 162ff, *165*
Mycetozoa, 4, 5
Myxobacteria, 14
myxomycetes, 3, 4, *6*, 8ff

Nägler, K., 25
Neely, C. L., Jr., 118, 120
Neurospora, germination of, 80
neutral red, 59
Nile blue sulphate, 59, 97
Noël, E., 140
nuclei, size of, 48, 51, 59

Oehler, R., 133
Olive, E. W., 19, 21, 23, 24, 26, 36,
  40, 47, 66, 73, 105, 157
Olive, L. S., 11, 14, 19ff, 26, 28ff,
  47, 54, 73, 85
Osborn, M. J., 130

PAS reaction, 34, 61, 68, 72, 155,
  159

· 203 ·

INDEX

of *Guttulina*, 28; of *Guttulinopsis*, 28; of *Polysphondylium*, 70
stalk formation, cell death in, 67, 169; cell sheath in, 66; interruption in, 43, 63; in *Polysphondylium violaceum*, 138
starvation, 133
Steinberg, M. S., 161, 162
Stoianovitch, C., 14, 19, 21, 28ff, 85
succinic dehydrogenase, 50, 129
Sussman, A. S., 80
Sussman, M., 55, 84, 86, 109, 113, 130, 131, 139, 140, 147, 162, 163, 164, 166, 167, 169
Sussman, R. R., 55, 113, 130, 139, 140, 162
synergism, 164, 165

Takeuchi, I., 50, 60, 77, 84, 129, 131, 173
taper, of stalk, 34, 65
temperature, effect on aggregation, 133; effect on germination of spores, 78; effect on growth, 82; effect on spore differentiation, 69, 136, 137; gradients, orientation in, 120

Tewari, J. P., 73
Thaxter, R., 14
Thom, C., 43, 157, 158
Tieghem, P. van, 3, 21, 23, 24, 28, 36
Townes, P. L., 161
trehalase, 131
trehalose, 80, 81, 131

UDP-Gal polysaccharide transferase, 130

vital staining, 58, 59, 63, 97

Watson, S. W., 5
Weinkauff, A. M., 42, 138
Weiss, P., 110, 161
Wenrich, D. H., 25
White, G. J., 84, 130, 164
Whitfield, F. E., 120
Whittingham, W. F., 67, 69, 137
Wilson, C. M., 13, 26, 45, 47, 53, 54, 55, 77
Wilson, M., 9
Wright, B. E., 113, 129, 140

Zahler, S. A., 15